INTRODUCING
Mathematics

Ziauddin Sardar • Jerry Ravetz • Borin Van Loon

Edited by Richard Appignanesi

Icon Books UK Totem Books USA

This edition published in the UK in
2005 by Icon Books Ltd.,
The Old Dairy, Brook Road,
Thriplow, Royston SG8 7RG
email: info@iconbooks.co.uk
www.iconbooks.co.uk

Sold in the UK, Europe, South Africa
and Asia by Faber and Faber Ltd.,
3 Queen Square, London WC1N 3AU
or their agents

Distributed in the UK, Europe, South
Africa and Asia by TBS Ltd., Frating
Distribution Centre, Colchester Road,
Frating Green, Colchester CO7 7DW

This edition published in Australia in
2005 by Allen and Unwin Pty. Ltd.,
PO Box 8500, 83 Alexander Street,
Crows Nest, NSW 2065

Original edition published in 1999

Reprinted 1999, 2000, 2001, 2002

This edition published in the USA in
2005 by Totem Books
Inquiries to: Icon Books Ltd.,
The Old Dairy, Brook Road,
Thriplow, Royston
SG8 7RG, UK

Distributed to the trade in the USA by
National Book Network Inc.,
4720 Boston Way, Lanham,
Maryland 20706

Distributed in Canada by
Penguin Books Canada,
10 Alcorn Avenue, Suite 300,
Toronto, Ontario M4V 3B2

ISBN 1 84046 637 5

Originating editor: Richard Appignanesi

Printed and bound in Singapore
by Tien Wah Press Ltd.

Why Maths?

Everybody moans at the very mention of "maths". People think that the world is divided into two kinds of folks. The "brainy" lot who understand mathematics but are not the kind of people one wants to meet at parties...

...and the rest of us!
Just keep your eyes open for any mathematicians, O.K.?
JERRY
ZIA

But all of us need to understand maths to some extent. Without mathematics, life would be inconceivable.

We need maths when we go shopping, check our bills, manage household finances...
...and run our businesses.
We need maths to build our houses ...
...insure our cars, do our banking.
We need maths to make maps so we can find our way around cities ...
...travel around the world, even go out into space!

Thus, mathematics is the engine that runs our industrial civilization.
It is the language of science, technology and engineering.
It is essential for architecture and design as well as economics and medicine.
Even art relies on mathematics to some extent.

Indeed, mathematics has become a guide to the world in which we live, the world which we shape and change, and of which we are a part. And as the world becomes more and more complex, and uncertainties in our environment become more urgent and threatening, we need mathematics to describe the risks we face and to plan our remedies.

The ability to deal with mathematics does require a special talent and skill – like any other field of human endeavour, such as dancing. Just as an accomplished ballet performance is sophisticated and exquisite, so is mathematics in its essence very elegant and beautiful.

But even though most of us cannot become fully-fledged ballet performers, all of us know what it is to dance and virtually all of us can dance. Similarly, all of us should know what mathematics is about, and be able to understand and handle certain basic steps.

Fear of maths is like fear of dancing.
Both are overcome by a little bit of practice.
Music is the pleasure the human soul experiences from counting without being aware that it is counting.
LEIBNIZ

To some extent, young beginners at mathematics retrace the steps of humanity in the development of mathematical knowledge.
If there isn't such a thing, then what is there at the end?

Counting

At school, children learn to count, to calculate, and to measure. Once they have been learned, these techniques may seem "elementary". But for the learners they are full of mystery.

The naming of numbers becomes an incantation, especially when we get to the bigger ones. Counting to a hundred becomes tedious, but getting to a thousand is like climbing a mountain! What is the last number, the biggest one of all?

How do we name the numbers, as we call them out one after another? Perhaps just a few numbers are enough. Some animals can recognize different collections up to five or seven – beyond that it's just "many". But if we know that numbers go on continuously, we can't just keep inventing new names indefinitely as we go along.

The language of the Dakota Indians was not written down.

So we counted the years and marked special events in our history by keeping a winter count like this one.

It is made of cloth and the pictographs are drawn in black ink. Each year a new pictograph was added to show the main event of the past year.

The best way to systematize naming and counting is to have a "**base**", a number that marks the beginning of counting again. The simplest base is just two. For example, the Gumulgal, an Australian indigenous people, counted like this:

1 = *urapon*
2 = *ukasar*
3 = *urapon-ukasar*
4 = *ukasar-ukasar*
5 = *ukasar-ukasar-urapon*

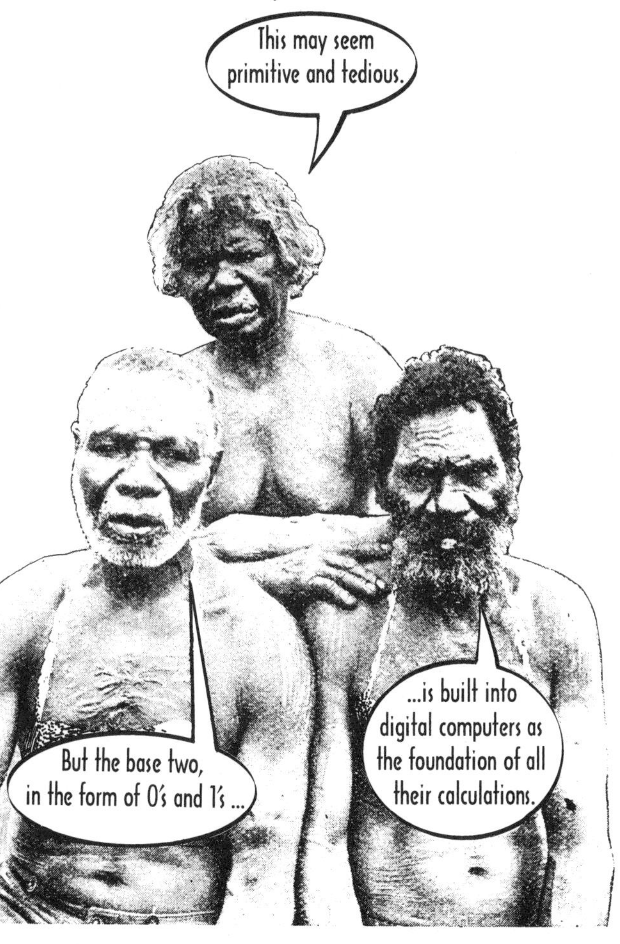
This may seem primitive and tedious.
But the base two, in the form of 0's and 1's ...
...is built into digital computers as the foundation of all their calculations.

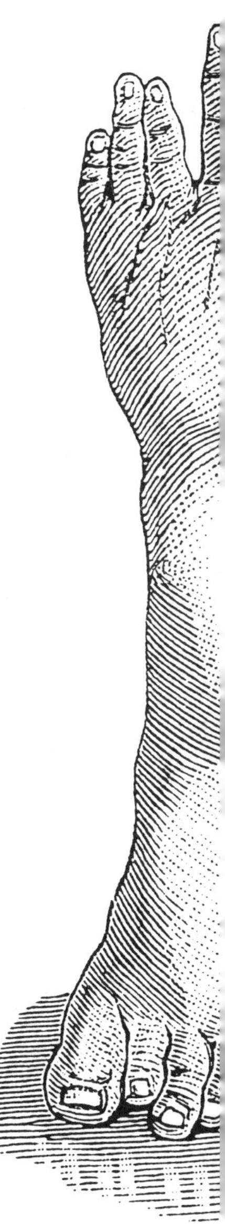

0
1
2
3
4
5

The fingers of the hands are useful for defining bases. Some systems use five, more common is ten. But many other bases can be used. The old British currency had several: twelve (pence per shilling) and then twenty (shillings per pound) and even twenty-one (shillings per guinea!). Shop assistants needed to keep reckoning books by their sides. And when people bought in instalments, they might be told that their living-room suite cost 155 guineas, or 104 weekly payments of one pound, fifteen shillings and sevenpence-halfpenny.

Who could calculate the interest on that?
Small wonder that instalment payments were called the "never-never" –
– you never finish paying!

The base twenty (fingers and toes?) is also common. The Yoruba used this, employing subtraction for the larger numbers within the base. They had different names for the numbers one (*okan*) to ten (*eewa*). From eleven to fourteen, they simply added. So eleven became "one more than ten", and fourteen "four more than ten". But from fifteen onwards they subtracted. So fifteen became "twenty less five" and nineteen became "twenty less one". The base twenty still survives in French, where eighty is "four-twenties", and ninety-nine is "four-twenties-nineteen".

Those who deal with computers use bases built on two.

So no single base is "best". We can think of a number system as **designed** with different attributes: easy to remember, convenient in naming, useful for calculating, etc.

Once the grouping, or base, for a number system had been developed, it allowed the four basic functions of arithmetic...
ADDITION
SUBTRACTION
MULTIPLICATION
DIVISION
...to be developed easily.

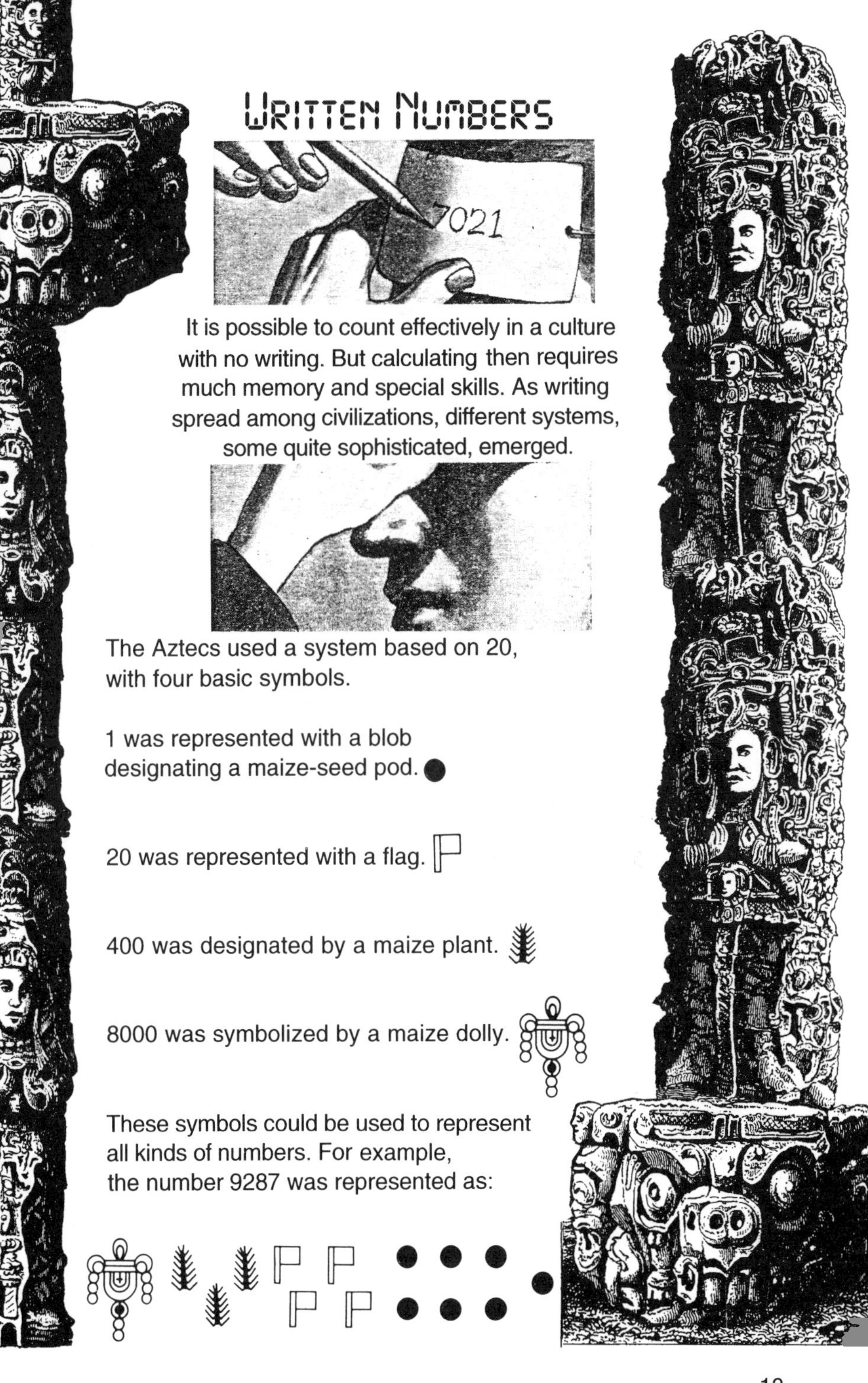
7021

Written Numbers

It is possible to count effectively in a culture with no writing. But calculating then requires much memory and special skills. As writing spread among civilizations, different systems, some quite sophisticated, emerged.

The Aztecs used a system based on 20, with four basic symbols.

1 was represented with a blob designating a maize-seed pod.

20 was represented with a flag.

400 was designated by a maize plant.

8000 was symbolized by a maize dolly.

These symbols could be used to represent all kinds of numbers. For example, the number 9287 was represented as:

0
1
2
3
4
5
6
7
8
9
10
11
12
13
14
15
16
17
18
...a large dot • was one,
...a bar ——— was five,
....and a snail's shell was zero.

The Mayans' numbering system had only three symbols:

So:

••• is 3

••• is 13

And twenty was represented by

The Ancient Egyptians (c. 4000-3000 BC) used a pictorial script (hieroglyph) to write down their numbers.

The pictograms, starting with one, increased by ten times, eventually reaching ten million.

1 10 100 1000 10,000 100,000 1,000,000 10,000,000

The Babylonians (c. 2000 BC) used a system based on 60 and its multiples, with the following symbols:

1 10 60 600 3600

Later, they evolved a system based on only two values:

for 1 (or 60 depending on its position) and for 10

So, 95 would be written as

$$95 = 60(1) + 35:$$

You know, I've lost count of the number of my wives...
Yes... As a Babylonian, I was able to spend nearly a whole extra hour in bed this morning...
I'll go to the foot of our stairs!

The Babylonian sexagesimal system has survived to this day. Circles have 360 degrees. Hours have sixty minutes. Minutes have sixty seconds.

The Ancient Chinese (c. 1400-1100 BC) used a base 10 system of numbers with symbols for one to ten, a hundred, a thousand and ten thousand. Later, around the 3rd century BC, the Chinese developed a form of numerals using straight lines (or rods),

So! It's a typical Oriental stereotype.

representing one to nine, which could be placed either upright:

or horizontally:

Normally, the uprights were used for units and hundreds and horizontals for tens and thousands. So, 6708 would be written as

with the blank space standing for "zero".

The Chinese made the great invention that put written symbols in a different world from the spoken names of numbers. This was a system of "place-value". The meaning of a number, as an expression of quantity, depended on its place in the string of numbers. Thus "2" could mean two, twenty, or two hundred, depending on its location. This made it unnecessary to name the higher bases – in "234" we know that the 2 means 200.

Elementary, my dear Watson. The number 2.689 is shown here with each figure proportionately-sized to show the quantity it represents! Hence the "value" of "place"...
2.689
20
6
.8
.09
The figure "9" is so small, I'm needed to make it readable!

The Indians developed three distinct types of number systems.

The Kharosthi (c. 400-200 BC) used symbols for ten and twenty, and numbers up to one hundred were built up by addition.

The Brahmi (c. 300 BC) used separate symbols for the digits one, four, to nine and ten, a hundred, a thousand and so on.

The Gwalior (c. 850 AD) had symbols for numbers one to nine as well as for zero.

Think of a number... O.K., now double it... treble it... quadrupeddle it...

The Indians were very comfortable with high numbers. The classical Hindu texts give names to numbers as large as 1,000,000,000,000 (*parardha*)!

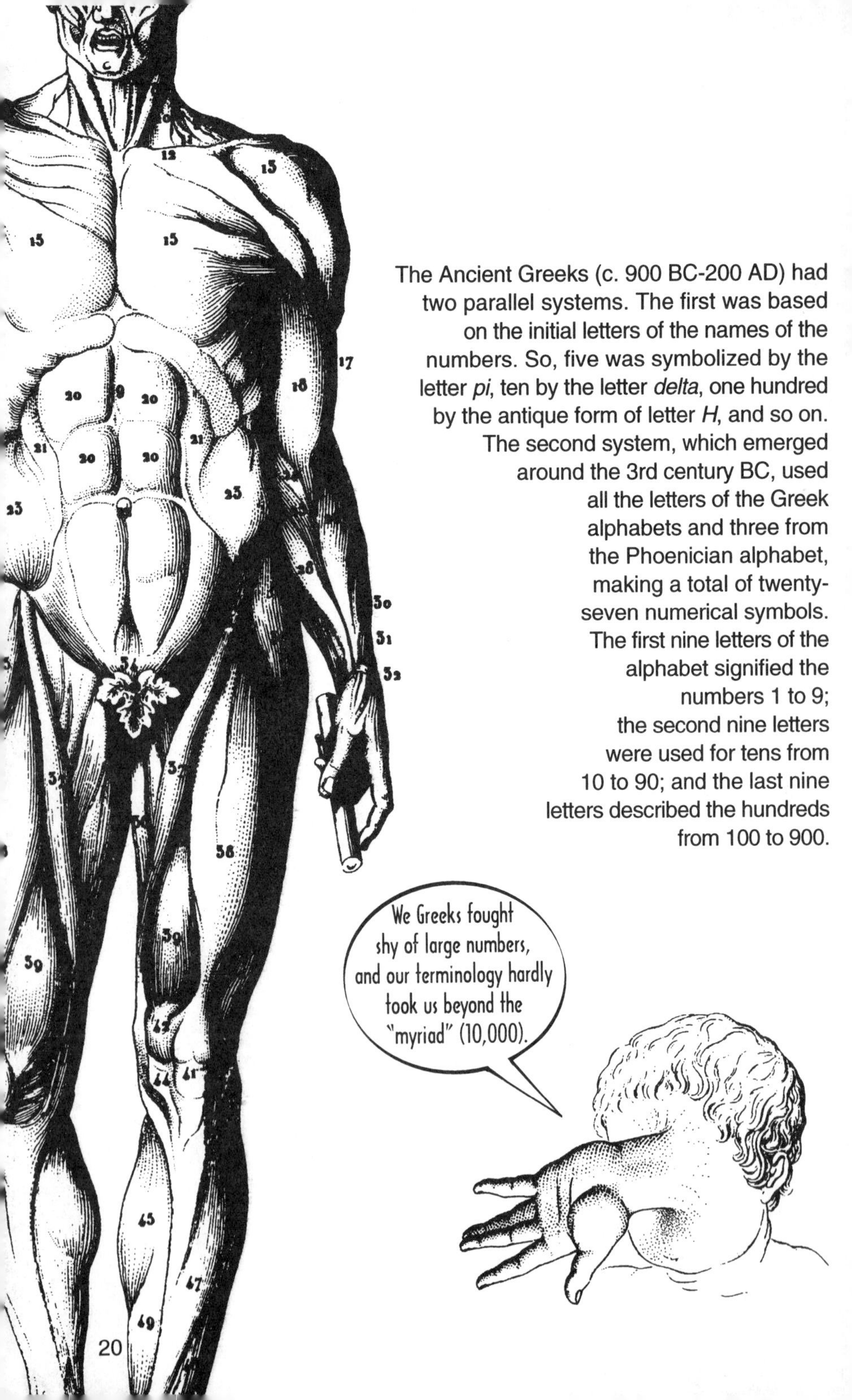
We Greeks fought shy of large numbers, and our terminology hardly took us beyond the "myriad" (10,000).

The Ancient Greeks (c. 900 BC-200 AD) had two parallel systems. The first was based on the initial letters of the names of the numbers. So, five was symbolized by the letter *pi*, ten by the letter *delta*, one hundred by the antique form of letter *H*, and so on. The second system, which emerged around the 3rd century BC, used all the letters of the Greek alphabets and three from the Phoenician alphabet, making a total of twenty-seven numerical symbols. The first nine letters of the alphabet signified the numbers 1 to 9; the second nine letters were used for tens from 10 to 90; and the last nine letters described the hundreds from 100 to 900.

The Roman system (400 BC-600 AD) had a total of seven symbols: I for 1, V for 5, X for 10, L for 50, C for 100, D for 500 and M for 1000.

The numbers are written from left to right with the largest quantities placed at the left and added together to obtain the designated number.

So, LX is 60.

For convenience, a smaller quantity on the left was interpreted as a subtraction. So, MCM means 1900.

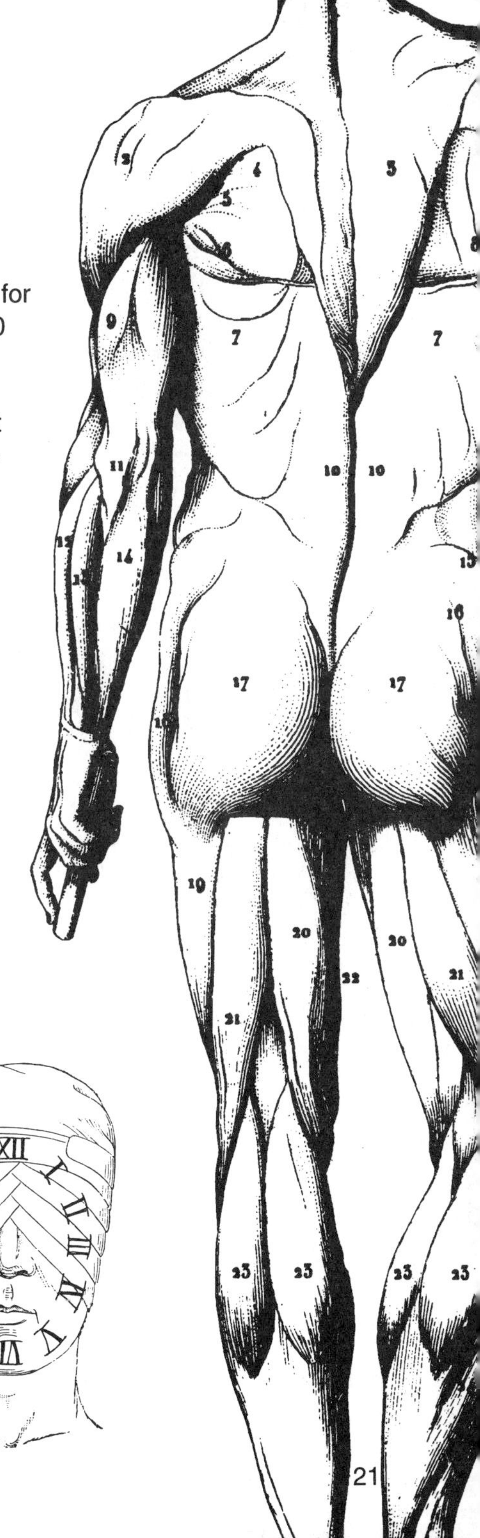

The Roman numerals, still used today for ornament, were not suitable for doing rapid calculations.

Oi!
Clock-face!

The use of the alphabet for numbers enabled the rise of a highly developed art of divination called "gematria". Given any word, or particularly a name, one would rearrange the letters to form a number and then scrutinize that for its quality and meaning. Anyone whose name yielded 666 (the Biblical "number of the beast") was obviously a Bad Thing!

Only with me, Descartes, and my successors in Europe did mathematics become completely "disenchanted", at least for the educated elite.
Now stop all this nonsense, or I'll put a curse on you.
Bad news, good young sirrah! Your name has the number "Spawn Of Satan."
I have the remedy here in my hand, my lady.
It was said that in World War II, resistance against me among some fundamentalist Christians was strengthened by the discovery that I was a 666 type.
Are you looking at my pint?

The Muslim civilization (650 AD-present) developed two sets of numerals. The sets were similar, but one was used in the eastern part of the Muslim world (Arabia and Persia), while the other was common in the western part (the Maghrib and Muslim Spain). Both contained ten symbols from zero to nine.

Eastern set: ٠ ٩ ٨ ٧ ٦ ٥ ٤ ٣ ٢ ١

Western set: 1 2 3 4 5 6 7 8 9 0

The eastern set is still used throughout the Arab world. The western set is what we now know as “Arabic numerals” – the system we all use today.

The next bus is a 971, I think.
But surely, it's a 250, is it not?
Don't ask me, I don't do all this new-fangled Western stuff.

The Zero

The Zero is a relatively late invention (about 6th century AD), and it seems to have been a joint product of the Chinese and Hindu civilizations. The Chinese needed some such thing for their place-value notation – how would they represent the missing place for the number "two hundred and five"? Just 25 is wrong, so they had something to "fill" the empty place, as 2–5. But the full meaning of zero was developed in the Indian civilization, where philosophical speculation on The Void was highly developed.

The answer's a big nothing.

That sort of cultural background was very necessary for the invention, for the Zero is very peculiar. In some ways it behaves like other numbers, for we can add with it.

0
But then multiplying zero times anything gives zero. It is possible to make paradoxes by using an equation like 2 x 0 = 4 x 0, and then cancelling the zero to get 2 = 4.
And what do we get...
....when we divide anything by zero?
Infinity!
FOSTER
THE BEAK

While zero is essential for calculation, it is excluded from counting. The first in a row of things is not the "zero-th". This paradox shows up in the calendar: the 1900s are the 20th century, since there was no zero-th century at the start of the Western calendar A.D.

Also, zero has two meanings, as we see from “the fossils joke”. A museum guide talks to a school party:

This fossil bone is sixty-five million and four years old.
How do you know so exactly?
Well, when I came to this job I was told that it was 65,000,000 years old... and that was four years ago.
65,000,000
+ 4
= 65,000,004

Of course everyone saw that this was ridiculous, but one of the pupils did the sum...

...just as she had been taught at school! No one had told her that the six zeros after 65 were just “filler” digits, not “counters”. For them, we don’t just have 0 x 4 = 0; we also have 0 + 4 = 0! Perhaps it was the awareness of paradoxes such as these (from which pupils are now carefully shielded) that made earlier mathematicians suspicious of strange numbers like zero.

Special Numbers

Besides zero, there are other sorts of special numbers that we have to be familiar with.

Some of them are "numbers with personality", which may be considered as having magical properties. The numbers 3, 5, 7 and 13 are each, in their own way, special. There are also kinds of numbers defined by their arithmetical properties, which attracted interest.

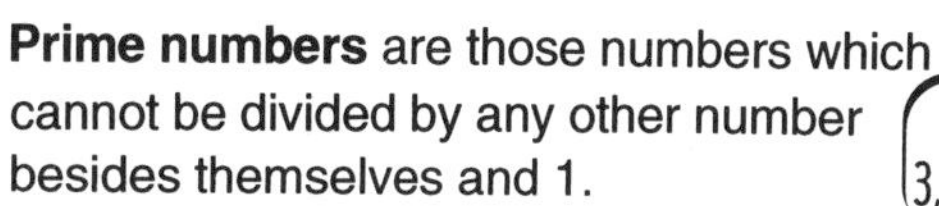

Prime numbers are those numbers which cannot be divided by any other number besides themselves and 1.

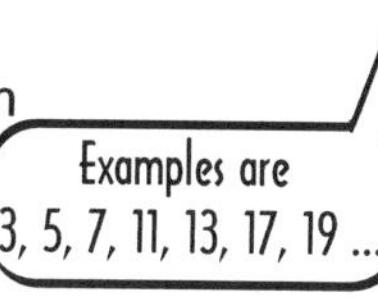

Examples are
3, 5, 7, 11, 13, 17, 19 ...

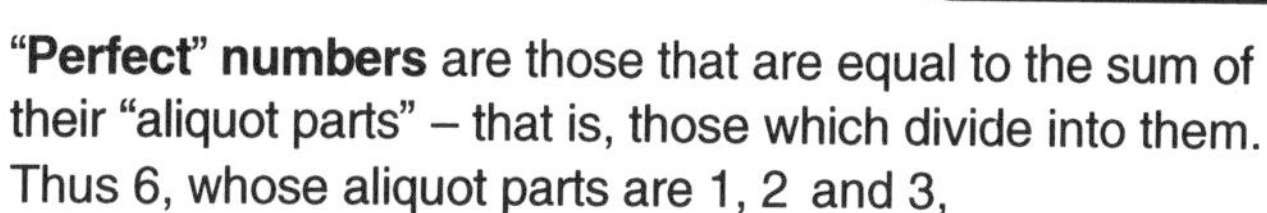

"Perfect" numbers are those that are equal to the sum of their "aliquot parts" – that is, those which divide into them. Thus 6, whose aliquot parts are 1, 2 and 3, is a perfect number, since 1 + 2 + 3 = 6.

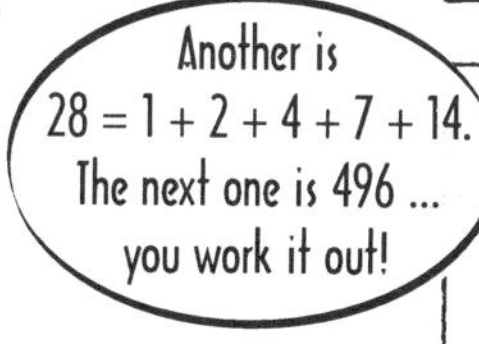

Another is 28 = 1 + 2 + 4 + 7 + 14. The next one is 496 ... you work it out!

Cutting the pie...
EXTRA!
1
2
4

8 isn't perfect...

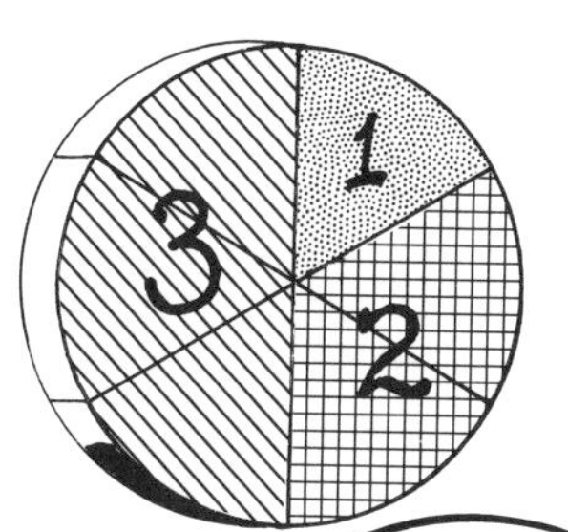

1
2
3

...but 6 is!

In ancient times, such numbers were considered very special – hence the name.

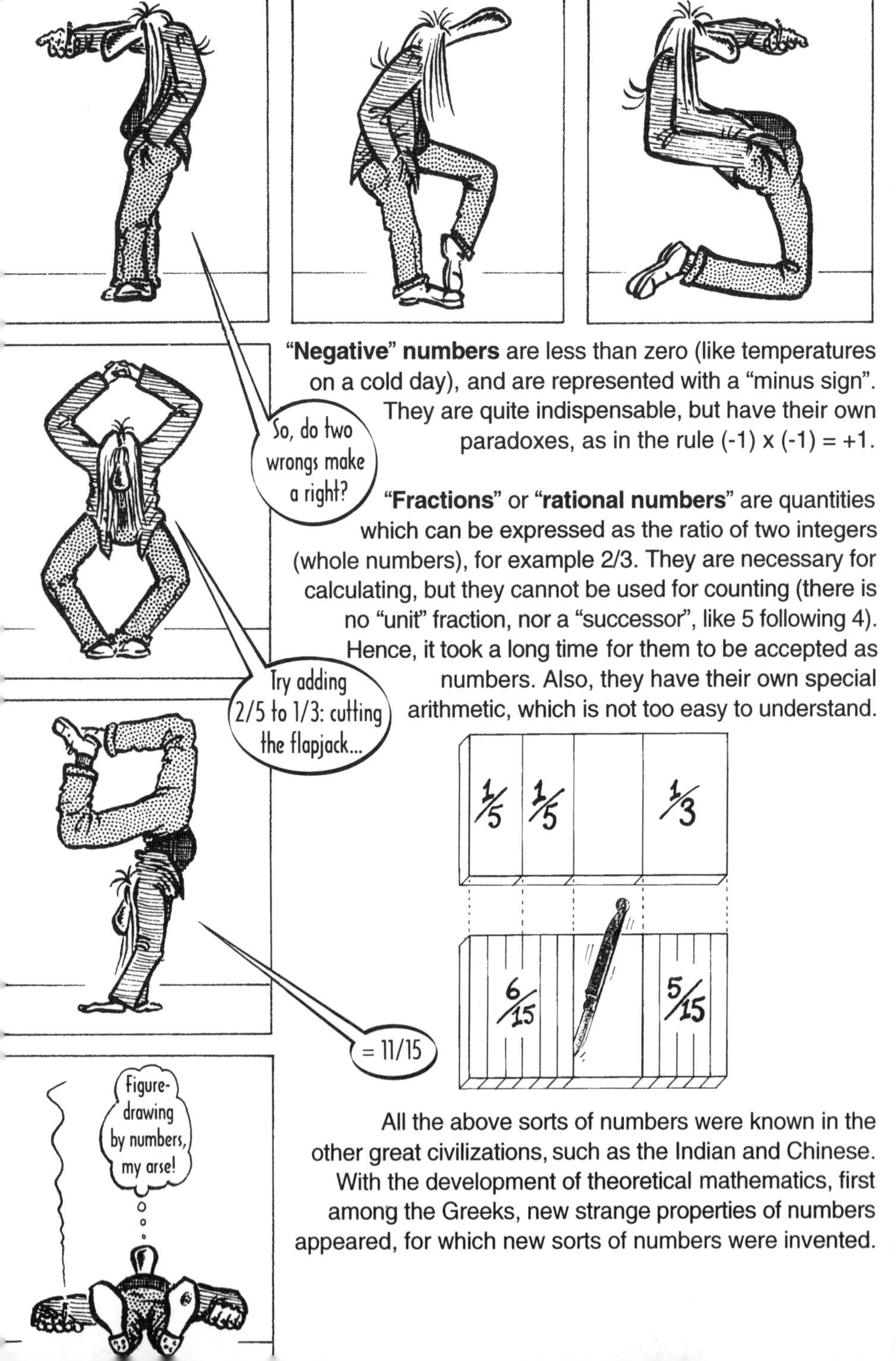
So, do two wrongs make a right?
Try adding 2/5 to 1/3: cutting the flapjack...
= 11/15
1/5 1/5 1/3
6/15 5/15
Figure-drawing by numbers, my arse!

"**Negative**" **numbers** are less than zero (like temperatures on a cold day), and are represented with a "minus sign". They are quite indispensable, but have their own paradoxes, as in the rule (-1) x (-1) = +1.

"**Fractions**" or "**rational numbers**" are quantities which can be expressed as the ratio of two integers (whole numbers), for example 2/3. They are necessary for calculating, but they cannot be used for counting (there is no "unit" fraction, nor a "successor", like 5 following 4). Hence, it took a long time for them to be accepted as numbers. Also, they have their own special arithmetic, which is not too easy to understand.

All the above sorts of numbers were known in the other great civilizations, such as the Indian and Chinese. With the development of theoretical mathematics, first among the Greeks, new strange properties of numbers appeared, for which new sorts of numbers were invented.

Irrational numbers are numbers which cannot be expressed as the ratio of two whole numbers. An important example is $\sqrt{2}$, produced by geometric operations. It is the length of the "hypotenuse" of a triangle with a right angle and equal sides of unit length.
Such numbers are called "surds".

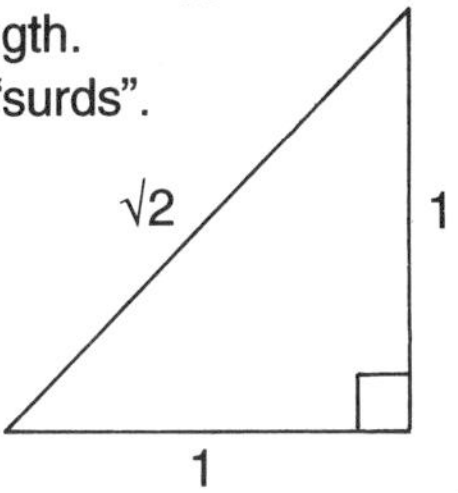

$\sqrt{2}$

1

1

Some quantities are very "irrational", being incapable of expression even by numbers produced by algebraic operations.

The most famous of these is "pi" or π, the ratio of the circumference of a circle to its diameter.

Reducing that ratio to surds was called the problem of "squaring the circle". Mathematicians tried for centuries, until in modern times it was shown to be impossible. Such numbers were then called...

..."transcendental".

'pi'...'pie', geddit?

Imaginary numbers are produced when real numbers are multiplied by the "imaginary" quantity, the square root of minus one (√-1).

When imaginary numbers are added to ordinary, or "real" numbers, the sum is called "**complex**".

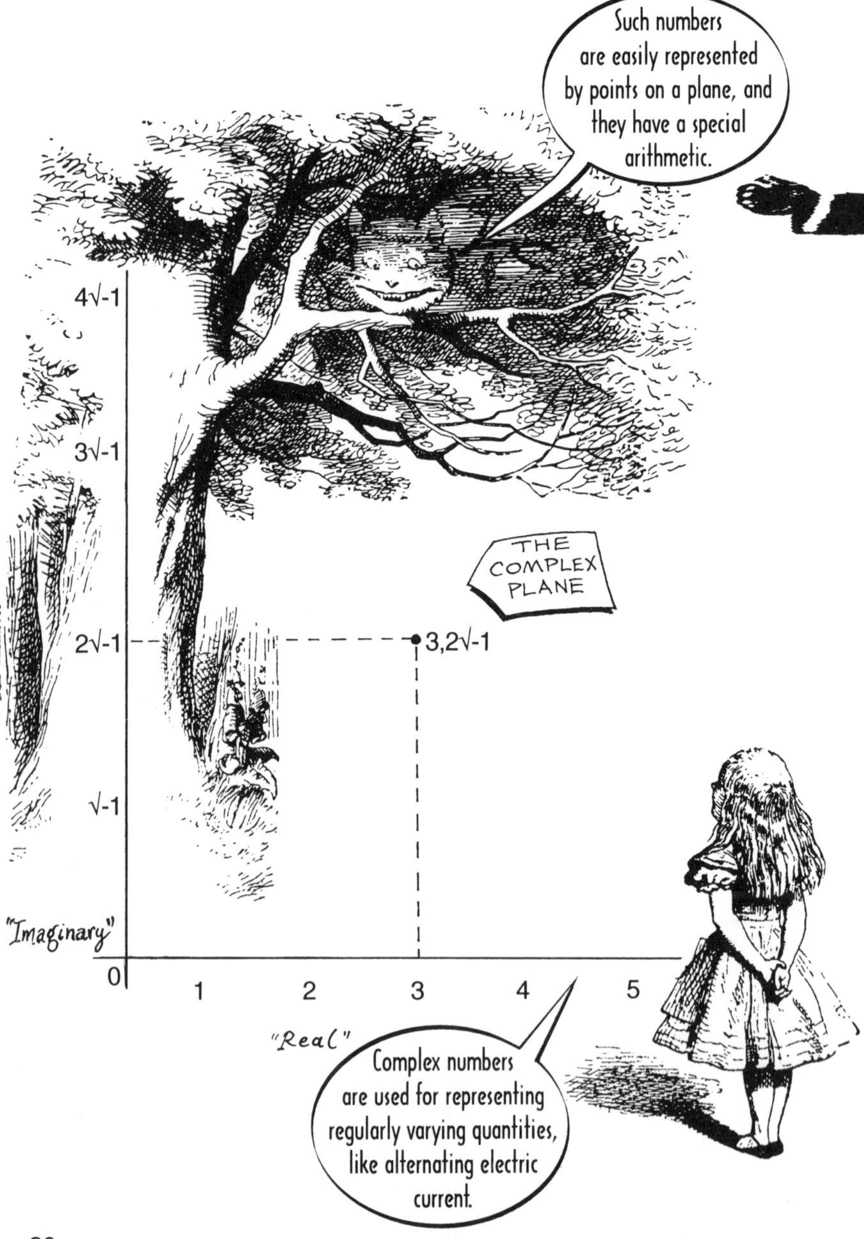

Such numbers are easily represented by points on a plane, and they have a special arithmetic.
4√-1
3√-1
2√-1
√-1
"Imaginary"
0
1
2
3
4
5
"Real"
3,2√-1
THE COMPLEX PLANE
Complex numbers are used for representing regularly varying quantities, like alternating electric current.

Large Numbers

Most of us are overawed by large numbers, and find it difficult to appreciate their real magnitude.

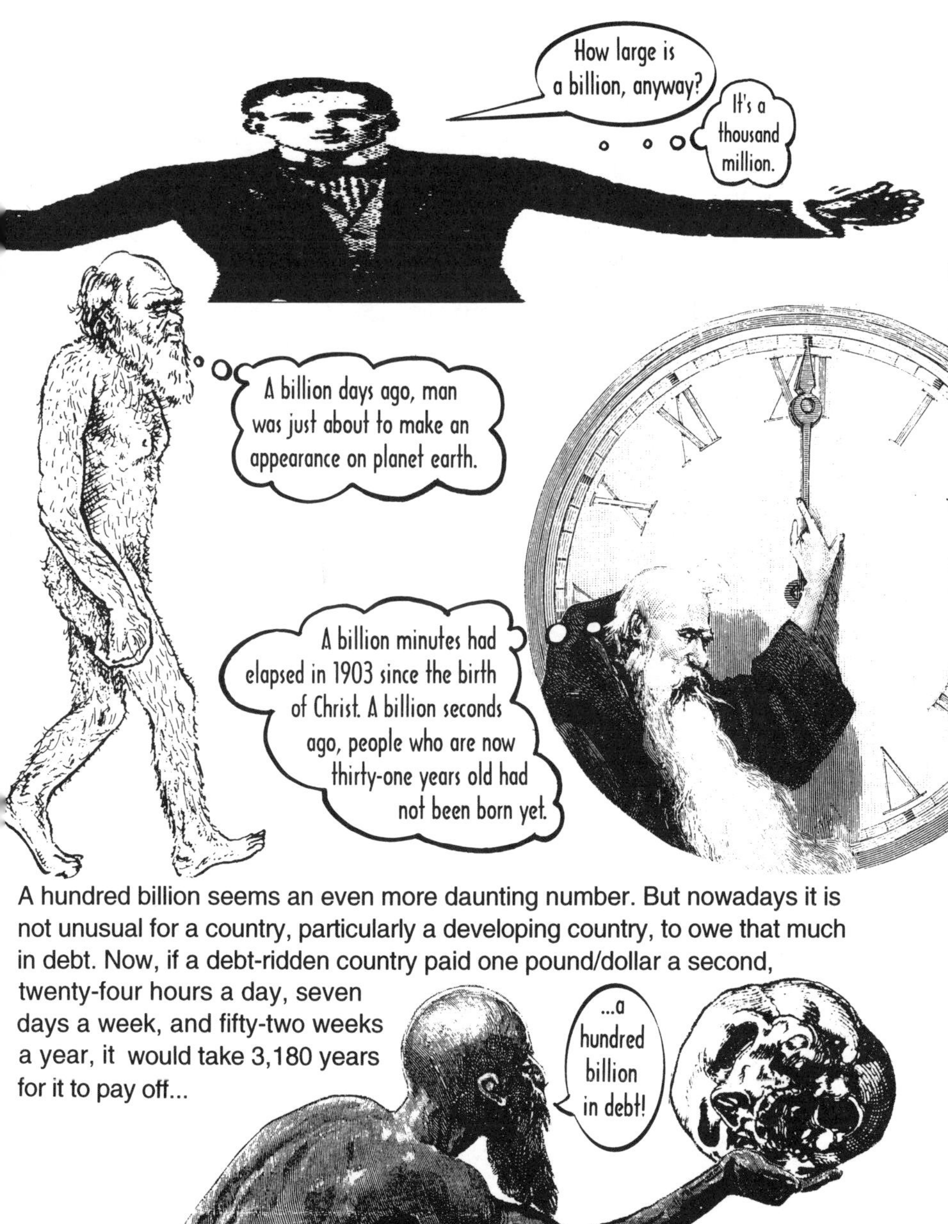

How large is a billion, anyway?
It's a thousand million.
A billion days ago, man was just about to make an appearance on planet earth.
A billion minutes had elapsed in 1903 since the birth of Christ. A billion seconds ago, people who are now thirty-one years old had not been born yet.

A hundred billion seems an even more daunting number. But nowadays it is not unusual for a country, particularly a developing country, to owe that much in debt. Now, if a debt-ridden country paid one pound/dollar a second, twenty-four hours a day, seven days a week, and fifty-two weeks a year, it would take 3,180 years for it to pay off...

...a hundred billion in debt!

Just how easily we can reach large numbers can be well illustrated by that old evil, the chain letter. A person sends a letter to two friends asking them each to copy the letter and send it to two of their friends, and so on. The first person sends 2 letters; at the second stage, 2 x 2 or 4 letters are sent; at the third stage, 2 x 2 x 2 or 8 letters are sent. How many sets will it take to reach a billion letters?

The thirtieth group of people along the line would alone send 1,073,741,824 letters!
But it never really happens! Eventually people run out of friends.

Powers

Great thunder! The powers are flooding into me!

It is quite cumbersome to write a billion: 1,000,000,000. Fortunately, there is a very convenient notion for writing large numbers. We can see that a billion is actually equal to:

$$10 \times 10 \times 10 \times 10 \times 10 \times 10 \times 10 \times 10 \times 10$$

So, if we denote the products of two 10's by 10^2 and three 10's by 10^3 and so on, we can write a million as 10^6 and a billion as 10^9. Moreover, we can write five billion as 5×10^9.

Raising anything to a power simply means multiplying it with itself as many times as indicated by the power. So 2^5 means $2 \times 2 \times 2 \times 2 \times 2$, or 32.

We can improve our familiarity with this notation by looking at the following problem.

What is the largest number that can be written with three 2's?
Here are the possibilities.

The smallest of these is $2^{2^2} = 2^4 = 16$. Then comes 222. Next $22^2 = 484$. The largest is $2^{22} = 4{,}194{,}304$.

The power notation also works for fractions. To turn a power into a fraction we simply place a negative sign in front of the power. So 10^{-1} means 1/10; 10^{-2} is 1/100; 10^{-3} is 1/1000, and so on.

Power relations represent the way things grow.
For example, if the distance between a projector and the screen is doubled, the picture on the screen does not take twice but four times as much space.
NO IT WASN'T
YES IT WAS
If the distance is trebled, the picture covers nine times its original space.

Similarly, if we enlarge a photograph or a map x times, x^2 times as much paper will be necessary.

x, x^2, x^3, x^4 and x^5 are called first, second, third, fourth and fifth powers of x. The earlier powers are described as "squared" and "cubed", from their geometrical meaning. Of course, instead of 2, 3, 4 or 5 we could have any number. Using n to stand for "any number", we say that x^n is called the nth power of x.

For a long time, mathematicians were puzzled by those higher powers; they could not imagine a hyper-space in which they could describe the shape.

In his book *The Dazzling*, written when he was only nineteen years old, the Muslim mathematician **ibn Yahya al-Samaw'al** (died c. 1175) first introduced the definition of . . .

The Power Of Zero
$x^0 = 1$
That is, anything raised to the power of zero is one.
Because if we multiply a thing by itself "no times at all", we get the unit.

Logarithms

A logarithm is the power to which one number must be raised in order to obtain another number. The first number is called the base. Since $10^2 = 100$, $\log_{10} 100 = 2$. We read this as: log to the base 10 of a hundred equals 2.

The most common bases for logarithms are 10 and the exponential e (see page 99).

As $x^0 = 1$ for any x, log 1 = 0 for all bases.

In order to multiply or divide two logarithmic expressions, we use the fact that multiplication and division of powers of a number corresponds to the addition and subtraction of these powers. So log (xXy) simply equals log x + log y.

Log-a-Rhythm Log-a-Music...
PTUI!
Addition is so much easier than multiplication.

Logarithms are a great asset in simplifying long and complicated calculations. To multiply (or divide) two numbers, you just look up their "logs" in a table, then add (or subtract), and finally locate the resulting number in the table and read off the sum (or quotient).

LOGARITHMS

	0	1	2	3	4	5	6	7	8	9	1	2	3	4	5	6	7	8	9
10	·0000	0043	0086	0128	0170	0212	0253	0294	0334	0374	4	8	12	17	21	25	29	33	37
11	·0414	0453	0492	0531	0569	0607	0645	0682	0719	0755	4	8	11	15	19	23	26	30	34
12	·0792	0828	0864	0899	0934	0969	1004	1038	1072	1106	3	7	10	14	17	21	24	28	31
13	·1139	1173	1206	1239	1271	1303	1335	1367	1399	1430	3	6	10	13	16	19	23	26	29
14	·1461	1492	1523	1553	1584	1614	1644	1673	1703	1732	3	6	9	12	15	18	21	24	27
15	·1761	1790	1818	1847	1875	1903	[illegible]	1959	1987	2014	3	6	8	11	14	17	20	22	25
16	·2041	2068	2095	2122	2148	[illegible]	[illegible]	2227	2253	2279	3	5	8	11	13	16	18	21	24
17	·2304	2330	2355	2380	[illegible]	[illegible]	[illegible]	2480	2504	2529	2	5	7	10	12	15	17	20	22
18	·2553	2577	[illegible]	[illegible]	[illegible]	[illegible]	[illegible]	[illegible]	2742	2765	2	5	7	9	12	14	16	19	21
19	·2788	2810	[illegible]	[illegible]	[illegible]	[illegible]	[illegible]	2945	2967	2989	2	4	7	9	11	13	16	18	20
20	·3010	[illegible]	[illegible]	[illegible]	[illegible]	3118	3139	3160	3181	3201	2	4	6	8	11	13	15	17	19
21	·3222	[illegible]	[illegible]	3284	3304	3324	3345	3365	3385	3404	2	4	6	8	10	12	14	16	18
22	·3424	[illegible]	3464	3483	3502	[illegible]	3541	3560	3579	3598	2	4	6	8	10	12	14	[illegible]	[illegible]
23	·3617	[illegible]	3655	3674	[illegible]	[illegible]	3729	3747	3766	3784	2	4	6	7	9	11	[illegible]	[illegible]	[illegible]
24	·3802	3820	3838	[illegible]	[illegible]	[illegible]	3909	3927	3945	3962	2	4	5	7	[illegible]	[illegible]	[illegible]	[illegible]	[illegible]
25	·3979	3997	[illegible]	[illegible]	[illegible]	[illegible]	4082	4099	4116	4133	2	3	5	[illegible]	[illegible]	[illegible]	[illegible]	[illegible]	[illegible]
26	·4150	4166	[illegible]	[illegible]	[illegible]	4232	4249	4265	4281	4298	2	[illegible]	[illegible]	[illegible]	[illegible]	[illegible]	[illegible]	[illegible]	[illegible]
27	·4314	[illegible]	[illegible]	[illegible]	4378	4393	4409	4425	[illegible]	[illegible]	[illegible]	[illegible]	[illegible]	[illegible]	[illegible]	[illegible]	[illegible]	[illegible]	[illegible]
28	·4472	[illegible]	[illegible]	[illegible]	4533	4548	4564	4579	[illegible]	[illegible]	[illegible]	[illegible]	[illegible]	[illegible]	[illegible]	[illegible]	[illegible]	[illegible]	[illegible]
29	[illegible]	[illegible]	[illegible]	4669	4683	4698	4713	4728	[illegible]	[illegible]	[illegible]	[illegible]	[illegible]	[illegible]	[illegible]	[illegible]	10	11	12
30	·4771	[illegible]	4800	4814	4829	4843	4857	4871	[illegible]	[illegible]	[illegible]	[illegible]	[illegible]	[illegible]	7	8	9	11	12
31	[illegible]	4928	4942	4955	4969	4983	4997	5011	5024	[illegible]	[illegible]	[illegible]	[illegible]	5	6	8	9	10	12
32	·5051	5065	5079	5092	5105	5119	5132	5145	5159	[illegible]	[illegible]	[illegible]	4	5	6	8	9	10	11
33	·5185	5198	5211	5224	5237	5250	5263	5276	5289	[illegible]	[illegible]	[illegible]	[illegible]	[illegible]	[illegible]	[illegible]	[illegible]	[illegible]	[illegible]
34	·5315	5328	5340	5353	5366	5378	5391	5403	5416	5428	1	3	4	5	6	7	9	10	11
35	·5441	5453	54'5	5478	5490	5502	5514	5527	5539	5551	1	2	4	5	6	7	8	10	11
36	·5563	5575	5587	5599	5611	5623	5635	5647	5658	5670	1	2	4	5	6	7	8	9	10
37	·5682	5694	5705	5717	5729	5740	5752	5763	5775	5786	1	2	3	5	6	7	8	9	10
38	·5798	5809	5821	5832	5843	5855	5866	5877	5888	5899	1	2	3	5	6	7	8	9	10
39	·5911	5922	5933	5944	5955	5966	5977	5988	5999	6010	1	2	3	4	5	7	8	9	10
40	·6021	6031	6042	6053	6064	6075	6085	6096	6107	6117	1	2	3	4	5	6	8	9	10
41	·6128	6138	6149	6160	6170	6180	6191	6201	6212	6222	1	2	3	4	5	6	7	8	9
42	·6232	6243	6253	6263	6274	6284	6294	6304	6314	6325	1	2	3	4	5	6	7	8	9
43	·6335	6345	6355	6365	6375	6385	6395	6405	6415	6425	1	2	3	4	5	6	7	8	9
44	·6435	6444	6454	6464	6474	6484	6493	6503	6513	6522	1	2	3	4	5	6	7	8	9
45	·6532	6542	6551	6561	6571	6580	6590	6599	6609	6618	1	2	3	4	5	6	7	7	8
46	·6628	6637	6646	6656	6665	6675	6684	6693	6702	6712	1	2	3	4	5	6	6	7	8
47	·6721	6730	6739	6749	6758	6767	6776	6785	6794	6803	1	2	3	4	5	5	6	7	8
48	·6812	6821	6830	6839	6848	6857	6866	6875	6884	6893	1	2	3	4	4	5	6	7	8
49	·6902	6911	6920	6928	6937	6946	6955	6964	6972	6981	1	2	3	4	4	5	6	7	8
50	·6990	6998	7007	7016	7024	7033	7042	7050	7059	7067	1	2	3	3	4	5	6	7	8
51	·7076	7084	7093	7101	7110	7118	7126	7135	7143	7152	1	2	3	3	4	5	6	7	7
52	·7160	7168	7177	7185	7193	7202	7210	7218	7226	7235	1	2	2	3	4	5	6	6	7
53	·7243	7251	7259	7267	7275	7284	7292	7300	7308	7316	1	2	2	3	4	5	6	6	7
54	·7324	7332	7340	7348	7356	7364	7372	7380	7388	7396	1	2	2	3	4	5	[illegible]	[illegible]	[illegible]
	0	1	2	3	4	5	6	7	8	9	1	2	3	4	5	6	7	8	9

	0	1	2	3	4
55	·7404	7412	7419	7427	[illegible]
56	·7482	7490	7497	7505	[illegible]
57	·7559	7566	7574	7582	[illegible]
58	·7634	7642	7649	7657	[illegible]
59	·7709	7716	7723	7731	[illegible]
60	·7782	7789	7796	7803	[illegible]
61	·7853	7860	7868	7875	[illegible]
62	·7924	7931	7938	7945	[illegible]
63	·7993	8000	8007	8014	[illegible]
[illegible]	[illegible]	8069	8075	8082	[illegible]
65	·8129	[illegible]	8142	8149	[illegible]
66	·8195	[illegible]	8209	8215	[illegible]
[illegible]	·8261	[illegible]	8274	8280	[illegible]
[illegible]	[illegible]	8331	8338	8344	[illegible]
[illegible]	·8388	8395	8401	8407	[illegible]
[illegible]	·8451	8457	8463	8470	[illegible]
[illegible]	·8513	8519	8525	8531	[illegible]
[illegible]	·8573	8579	8585	8591	[illegible]
73	·8633	8639	8645	8651	[illegible]
74	·8692	8698	8704	8710	[illegible]
75	·8751	8756	8762	8768	[illegible]
76	·8808	8814	8820	[illegible]	[illegible]
77	·8865	8871	8876	[illegible]	[illegible]
78	·8921	8927	8932	[illegible]	[illegible]
79	·8976	8982	8987	[illegible]	[illegible]
80	·9031	9036	9042	[illegible]	[illegible]
81	·9085	9090	9096	[illegible]	[illegible]
82	·9138	9143	9149	[illegible]	[illegible]
83	·9191	9196	9201	[illegible]	[illegible]
84	·9243	9248	9253	[illegible]	[illegible]
85	·9294	9299	9304	[illegible]	[illegible]
86	·9345	9350	9355	[illegible]	[illegible]
87	·9395	9400	9405	[illegible]	[illegible]
83	·9445	9450	9455	[illegible]	[illegible]
89	·9494	9499	9504	[illegible]	[illegible]
90	·9542	9547	9552	[illegible]	[illegible]
91	·9590	9595	9600	[illegible]	[illegible]
92	·9638	9643	9647	[illegible]	[illegible]
93	·9685	9689	9694	[illegible]	[illegible]
94	·9731	9736	9741	[illegible]	[illegible]
95	·9777	9782	9786	[illegible]	[illegible]
96	·9823	9827	9832	[illegible]	[illegible]
97	·9868	9872	[illegible]	[illegible]	[illegible]
98	·9912	9917	[illegible]	[illegible]	[illegible]
99	·9956	9961	[illegible]	[illegible]	[illegible]
	0	1	2		

Log 2.2 = .3424

Log 3 = .4771

Add the Logs together to get .8195 which is log 6.6 (ie. 2.2 x 3)

I'll have to use my log-rule and slide-tables ...

The first tables were constructed by the Scottish mathematician **John Napier** (1550-1617). They were to the base e, and are called "natural" (because of the base) or "Napierian" (for the inventor).

Calculation

Manipulating numbers of all kinds to get an answer is the process we call calculation. All mathematical operations involve calculations.

Calculation was once done with stones. The Ancient Greeks used pebbles to count and to do their elementary calculations. The root of the English word "calculate" is the Latin *calculus*, meaning "a pebble".

Counted
Out

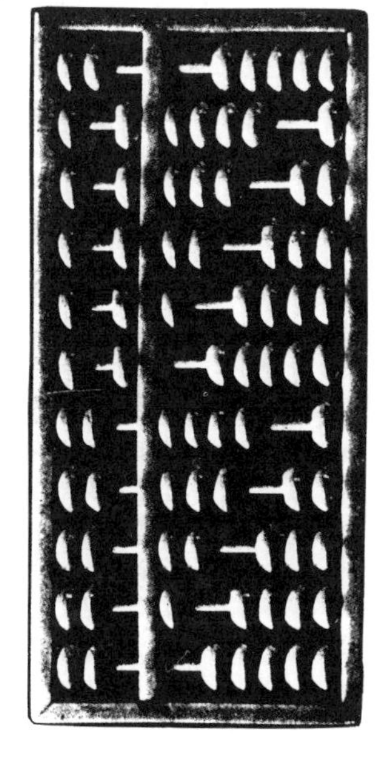

Until recently, the abacus, with beads on wires, was the most widespread reckoning device. Even today, a skilled abacus operator can manipulate the beads more quickly than a digital keyboard operator can find the keys!

Calculating devices come in two basic forms: simple adding machines limited to addition and subtraction, and calculators that not only perform multiplication and division...

...but numerous other functions as well.
PASCAL
LEIBNIZ
+/−
%
√
C·CE
R-CM
M−
M+
÷
7
8
9
×
4
5
6
−
1
2
3
+
0
•
=

The first adding machine was invented by the French mathematician **Blaise Pascal** (1623-62) in 1642, and was able to add and carry. In 1671, the German mathematician and philosopher **Gottfried Wilhelm von Leibniz** (1646-1716) produced a device that could multiply by repeated addition.

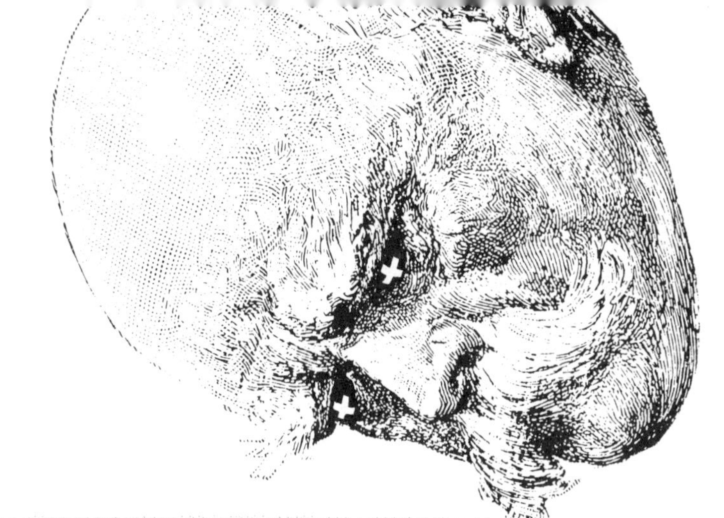

In 1822, the English mathematician and inventor **Charles Babbage** (1792-1871) built a small adding machine. Ten years later, he conceived his "Difference Engine", a predecessor of the digital computer. Then he embarked on a much more ambitious "Analytical Engine", which never got built. A replica of part of it has been built, however, and is in the Science Museum in London.

Calculations, however complicated, don't always suffice for solving problems. Sometimes equations are needed.

EQUATIONS

Equations are the core of mathematics. With the sole exception of very elementary mathematics, equations are used in all branches of pure and applied mathematics, as well as in physical, biological and social sciences.

As the name implies, an equation makes a statement of equality between two expressions. It usually involves quantities that are not known; in general, these will be called "**variables**", and others are "**constants**", or perhaps "**parameters**". Equations may also be used for defining quantities, or for expressing relationships among variables.

Constants
Variables
Variables
Knight?
When an equation is used to express the problem of finding the value of one of the variables, it will be called...
..."the unknown".

Before equations were invented, mathematical problems were solved by a variety of ingenious and complicated methods. Now they are reduced to a very simple form.

In the equation $5x + 8 = 23$, x is the unknown that has to be calculated. We can do it by trial and error, or by simple operations (subtracting 8 from both sides, then dividing both sides by 5).

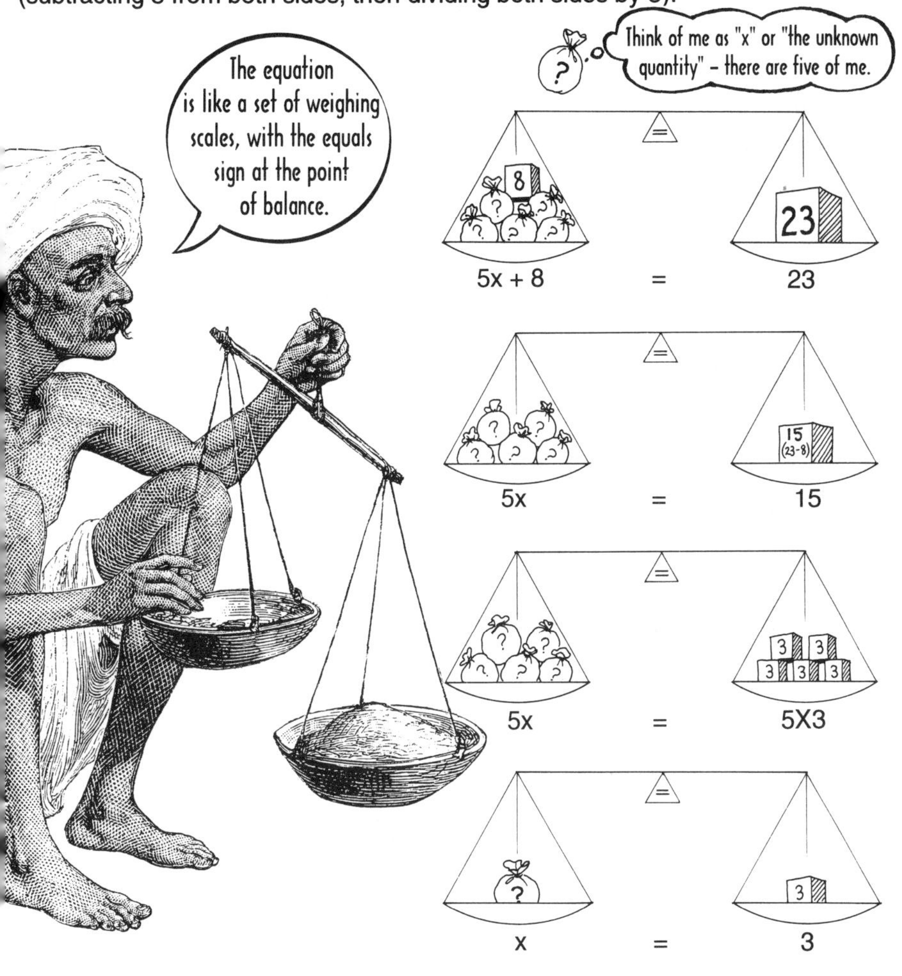

The equation is like a set of weighing scales, with the equals sign at the point of balance.
Think of me as "x" or "the unknown quantity" – there are five of me.
8
23
5x + 8
=
23
15
(23-8)
5x
=
15
3
5x
=
5X3
3
x
=
3

This equation is "satisfied" or "solved" when $x = 3$, as by checking we see that the two sides of the equation are then equal.

When all possible values of variables satisfy an equation, it is called an **identity**. For example, the equation $(x + y)^2 = x^2 + 2xy + y^2$ is an identity because it is true for all possible values of the unknown. Such identities are very useful for algebraic manipulation, as they enable a more complicated expression to be replaced by a simpler one.

Linear equations only have variables to the power of one – such as the equation $5x + 8 = 23$. They are called linear because when they are plotted on a graph they give a straight line.

Quadratic equations have a single variable that appears to the power 2. These equations always have two roots, although these roots could be equal. For example, $x^2 = 4$ and $2x^2 - 3x + 3 = 5$ are both quadratic equations. Their roots are, respectively, $(+2, -2)$ and $(2, -1/2)$. For equal roots, an example is $x^2 - 4x + 4 = 0$, with two roots, $x = 2$.

Cubic equations have a single variable to the power of 3. Cubic equations always have three roots, although two or all three of these roots may be equal, and two (but never all three!) may be complex. An example of a cubic equation is $x^3 - 6x^2 + 11x - 6 = 0$; its roots are $x = 1, 2, 3$.

Linear, quadratic and cubic equations are said to be of the first, second and third degrees respectively. Up to the quartic (fourth degree) equations, it is possible to express the roots of the equation by formulae involving arithmetic and square roots. Thus for the quadratic equation $ax^2 + bx + c = 0$, the formula is:

$x = [1/2a][-b \pm \sqrt{(b^2 - 4ac)}]$
The expression under the square root (√) sign may be less than zero; then we have a pair of "complex" roots.

There is no limit to the degree of such algebraic equations. But there is a break with the "quintic", or equation of fifth degree. For centuries, mathematicians tried to find a formula of arithmetic and square roots, like the one on page 45, for expressing the roots of the quintic. It was finally shown to be impossible in the early 19th century.

Equations can have more than one variable in each term. For example, the equation $xy = 1$ is a very basic one, describing the geometric figure, the "hyperbola".

x

o

y

Hyperbola
$xy = 1$

The **degree of an equation** is defined as the sum of the powers of the variables in the highest power term of the equation. For example, in the equation $ax^5 + bx^3y^3 + cx^2y^5 = 0$, the highest power term is cx^2y^5.

The sum of the powers in this term is 7, so the equation is of the 7th degree.
This time it's the third degree!
BIFF!
BOK!

Simultaneous equations involve two variables.
Oh no, double maths again...

A single equation with two variables is normally insoluble. But if there are as many equations as variables, it is possible to solve for each variable. A system of simultaneous equations has two or more equations involving two or more unknowns. These can sometimes be solved by simple manipulation.
For example:

1.

$$2x + xy + 3 = 0$$
$$x + 2xy = 0$$

2. If we multiply the first equation by 2, we get:

$$4x + 2xy + 6 = 0$$

3. And if we subtract the second equation from this, we get:

$$3x + 6 = 0$$

4. So we have: $x = -2$

Now, if we substitute this value into the first equation, we find the value $y = -1/2$.
More complicated simultaneous equations can sometimes be solved in the same way.

Most cannot be solved directly, and computers are used to obtain approximate solutions.
There are many other kinds of mathematical equations – such as trigonometric, logarithmic, differential and integral. We will meet them later.

Measurement

Measurements are an essential part of mathematics. We measure almost everything, from time to dimensions, weights to capacities, sizes to altitudes, electricity to heat and light – even the distance to stars and energies of sub-atomic particles. Now we even measure intelligence and the value of good things like the environment.

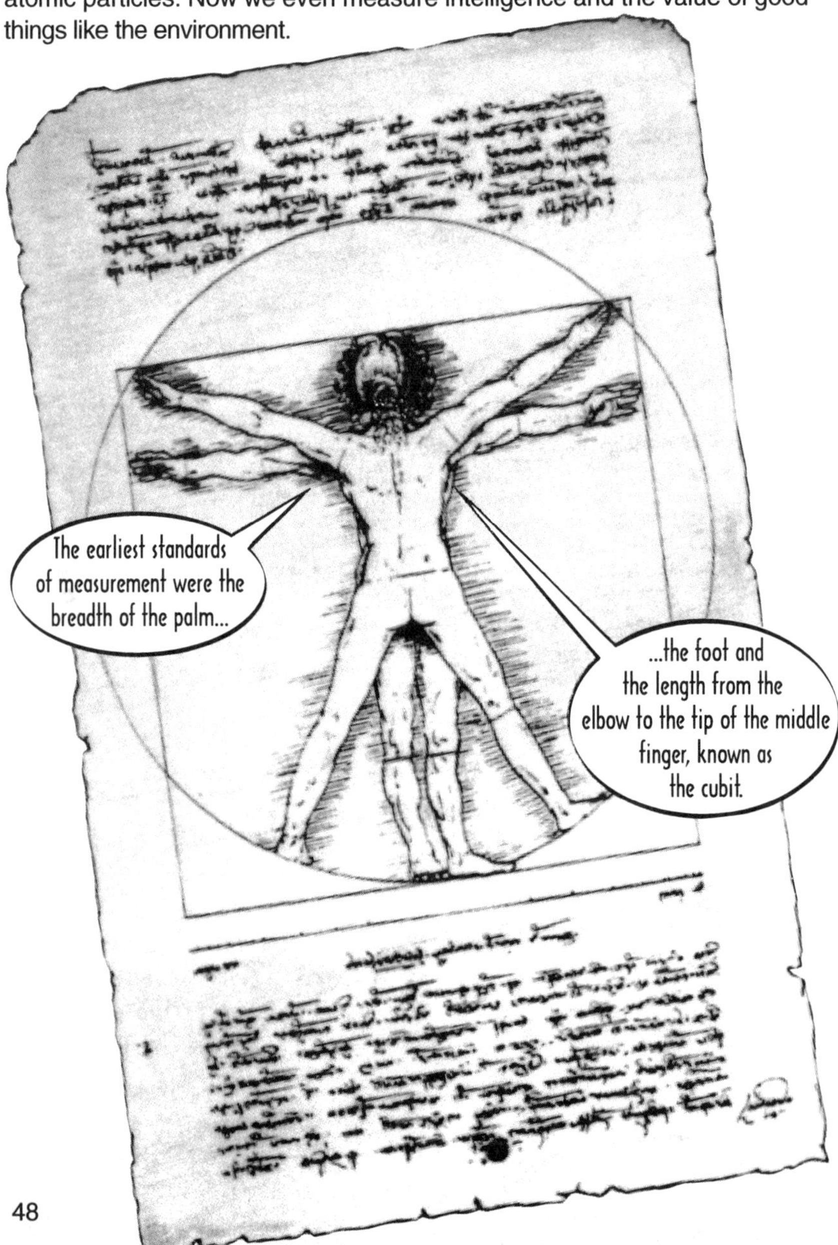

The earliest standards of measurement were the breadth of the palm...
...the foot and the length from the elbow to the tip of the middle finger, known as the cubit.

Now our measurements are based on science.
Time is an exception – the attempt by the French reformers to divide the month into three "decades" of ten days, and then the day into ten hours of a hundred minutes each, was very unpopular, and so we still use the system invented in Babylon.

The "Système Internationale" is descended from the "metric system" that was introduced during the French Revolution. It provides a connected set of units derived from basic quantities, such as the metre (m) for length, the second (s) for time, and the kilogram (kg) for mass. Most of the practical measures are expressed in powers of ten of the units, such as millimetre (mm) for length.

Each of the fundamental units has a definition and measurement procedures that are monitored by official international committees. Definitions are changed when better methods appear.

Many countries still use the old "Imperial" system involving pounds and yards, pints and quarts. But beware: the American pint, quart and gallon are only four-fifths as big as the English; so those "gas guzzling" cars with their low mileages to the gallon...

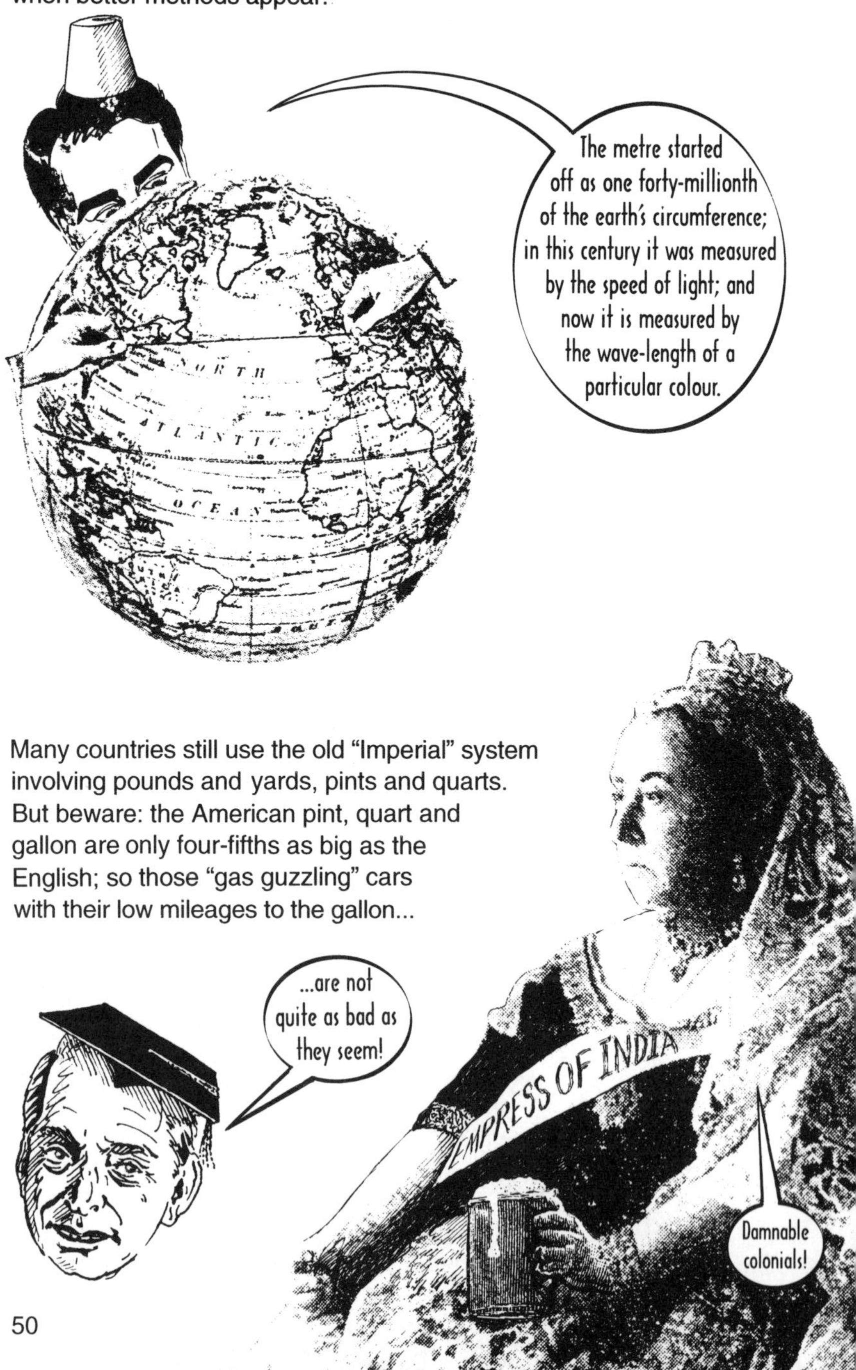
The metre started off as one forty-millionth of the earth's circumference; in this century it was measured by the speed of light; and now it is measured by the wave-length of a particular colour.
...are not quite as bad as they seem!
EMPRESS OF INDIA
Damnable colonials!

Counting and calculation concern separate, discrete quantities, involving exact numbers. Measurement, by contrast, concerns continuous magnitudes. No measurement is exact. When we compare the object being measured against a standard, we always interpolate between the points on the finest scale. And every report of a complex measurement has (or should have!) an "error bar" to indicate the "fringe" of uncertainty associated with it.

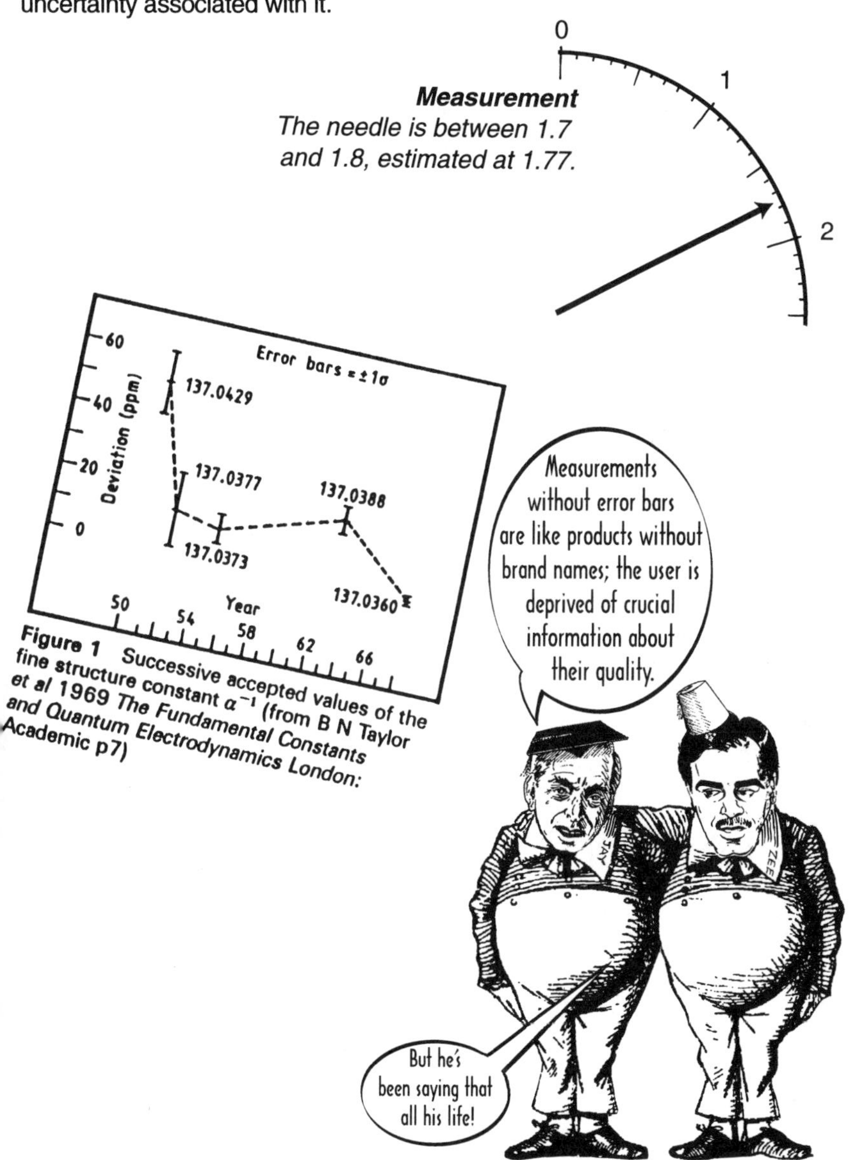
0
1
2
Measurement
The needle is between 1.7 and 1.8, estimated at 1.77.
60
40
20
0
Deviation (ppm)
Error bars = ±1σ
137.0429
137.0377
137.0388
137.0373
137.0360
50
54
58
62
66
Year
Measurements without error bars are like products without brand names; the user is deprived of crucial information about their quality.
JAY
ZEE
But he's been saying that all his life!

Figure 1 Successive accepted values of the fine structure constant α^{-1} (from B N Taylor *et al* 1969 *The Fundamental Constants and Quantum Electrodynamics* London: Academic p7)

Since prehistory, measurements have been used for building and design. Archaeologists have discovered that ancient monuments like Stonehenge were precisely aligned for the sighting of astronomical events, and their ground plans required geometrical constructions for their design. The churches of medieval Europe were designed in subtle proportions; and during the Renaissance, the theory of "the divine proportion" underlay architecture and art. The great pyramids of Egypt have provided a challenge for generations of archaeologists.

Were their proportions the expression of special magical numerical relations?
I still think it would benefit from a lean-to conservatory

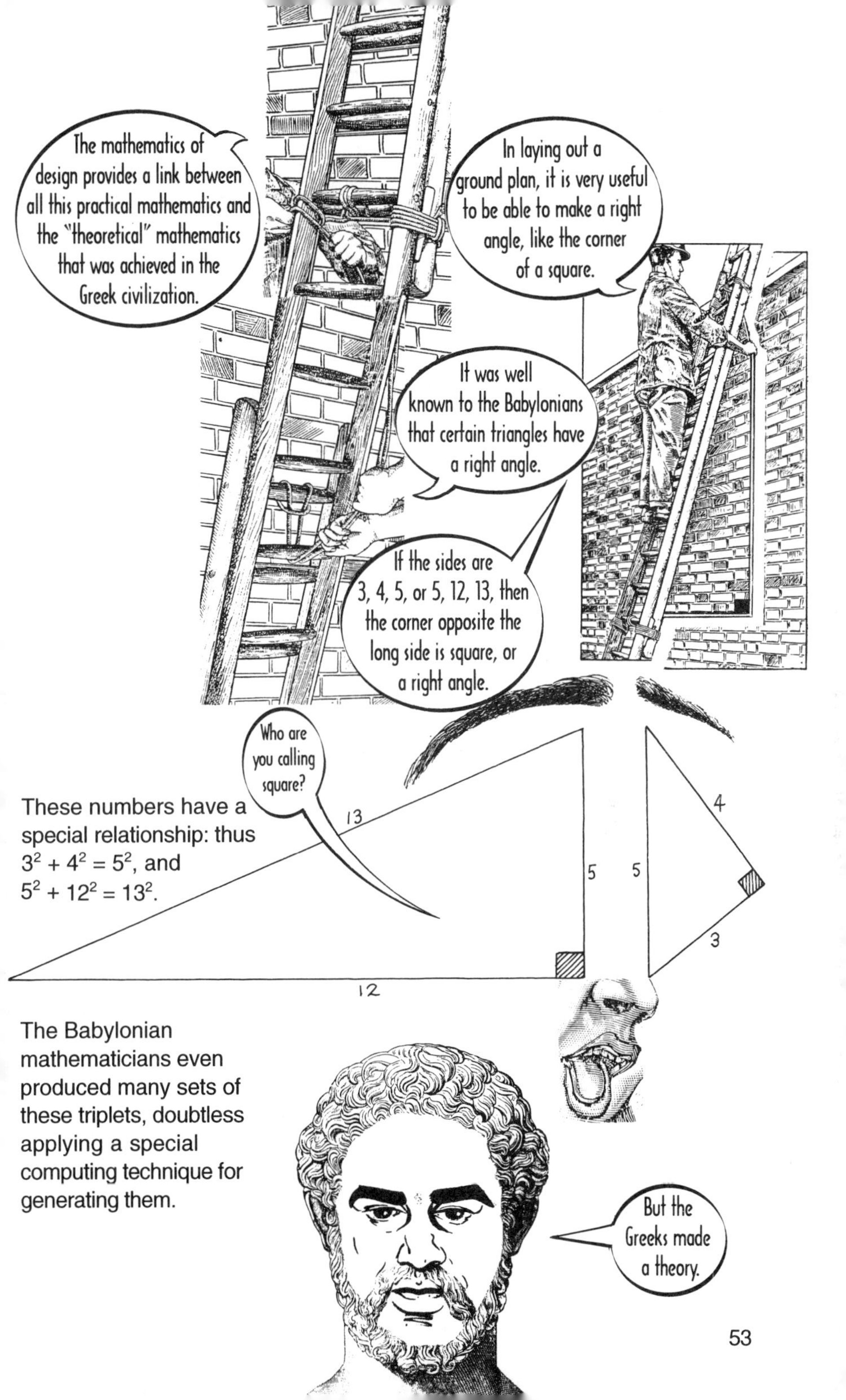
The mathematics of design provides a link between all this practical mathematics and the "theoretical" mathematics that was achieved in the Greek civilization.
In laying out a ground plan, it is very useful to be able to make a right angle, like the corner of a square.
It was well known to the Babylonians that certain triangles have a right angle.
If the sides are 3, 4, 5, or 5, 12, 13, then the corner opposite the long side is square, or a right angle.
Who are you calling square?
13
5
12
5
4
3
But the Greeks made a theory.

These numbers have a special relationship: thus $3^2 + 4^2 = 5^2$, and $5^2 + 12^2 = 13^2$.

The Babylonian mathematicians even produced many sets of these triplets, doubtless applying a special computing technique for generating them.

Greek Mathematics

From the 7th century BC onwards, the Greeks gradually separated the investigation of the laws of nature from religious questions of the relationship between man and the gods. **Thales of Miletos** (c. 624 BC), a statesman and mathematician, is said by Aristotle to have brought mathematics to Greece from Egypt.

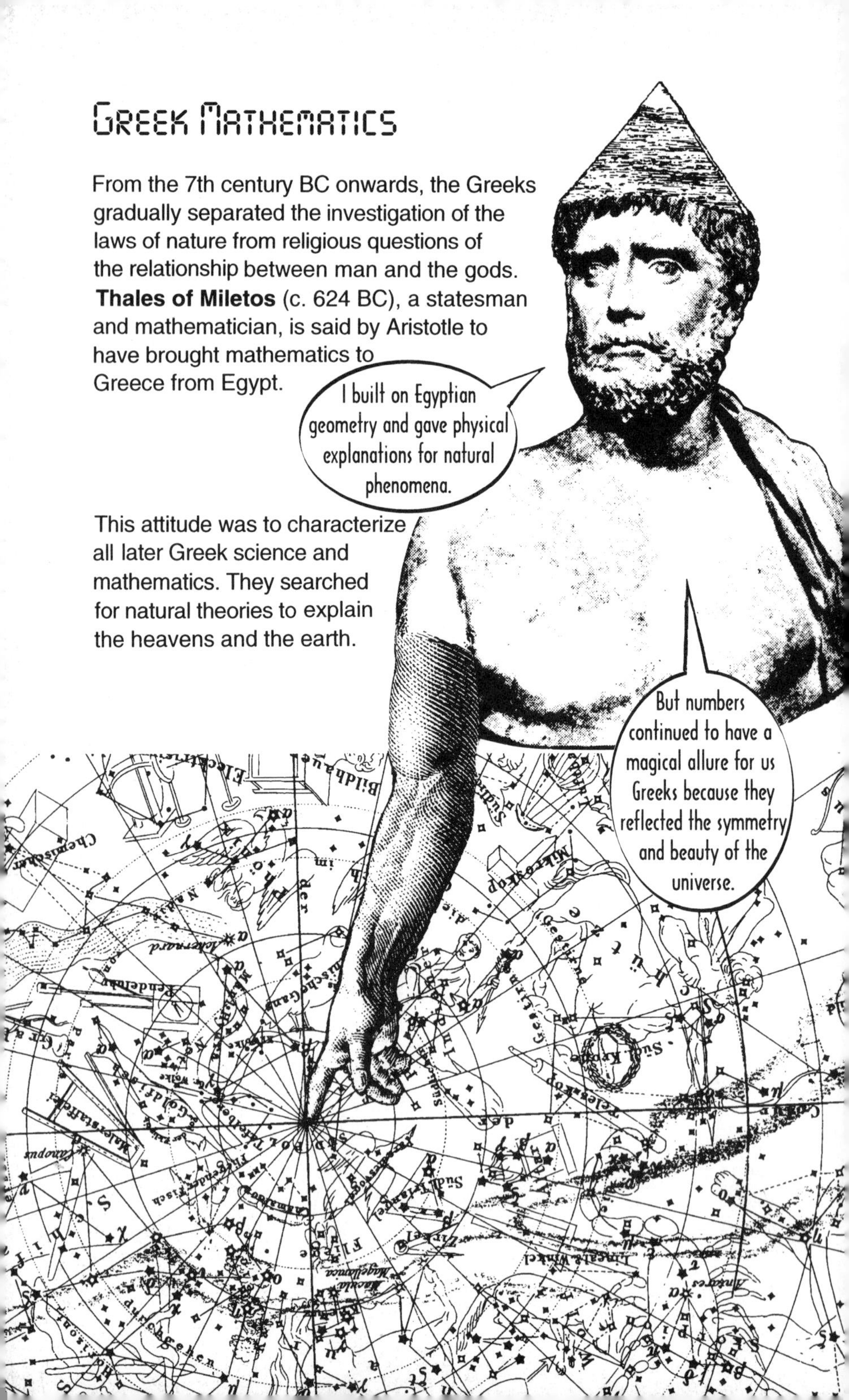

I built on Egyptian geometry and gave physical explanations for natural phenomena.
But numbers continued to have a magical allure for us Greeks because they reflected the symmetry and beauty of the universe.

This attitude was to characterize all later Greek science and mathematics. They searched for natural theories to explain the heavens and the earth.

Pythagoras

I, Pythagoras (580-500 BC) was not just a mathematician but also a civic leader and founder of a mystical cult that practised ascetic exercises and abstinence from various foods and activities.

The Pythagoreans had discovered that simple musical harmonies are made by the combination of instruments with simple ratios of lengths. The octave is made by two strings, one half the length of the other; for the major fifth, the ratio is 2 to 3.

This led us to believe that mathematics reflected beauty and divine relationships – numbers held the answers to everything and had a magical quality.

Pythagoras is credited with the famous theorem named after him, which says that in a right-angle triangle, the squares of the two sides are equal to the square of the hypotenuse, i.e. $a^2 + b^2 = c^2$. As we have seen, this was well known already, but we can suppose that Pythagoras was the first to attempt a general proof. Although this story did not appear for hundreds of years after his life, it does fit with what we know of his endeavour to change mathematics from being a merely practical study to one with philosophical significance.

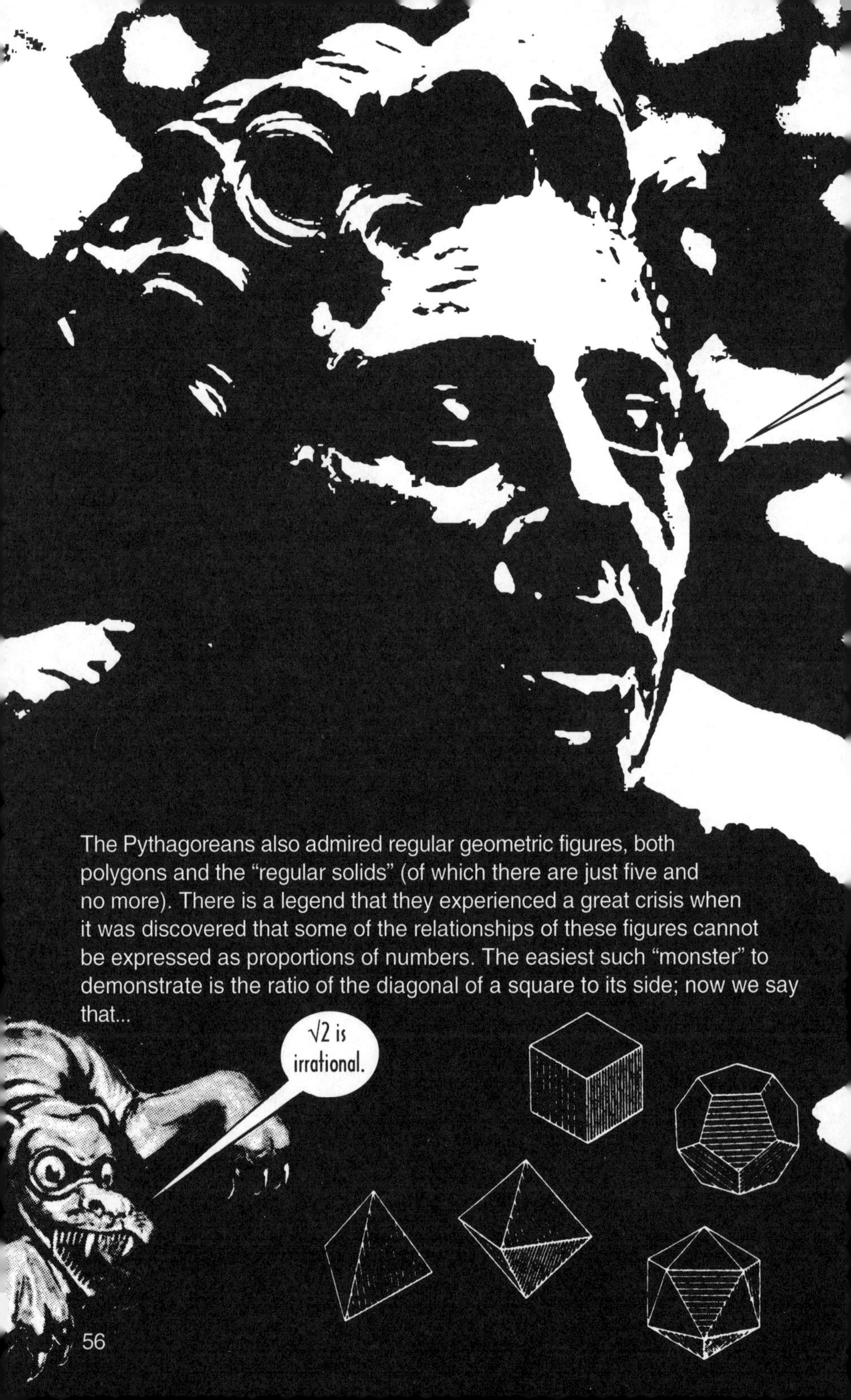

The Pythagoreans also admired regular geometric figures, both polygons and the "regular solids" (of which there are just five and no more). There is a legend that they experienced a great crisis when it was discovered that some of the relationships of these figures cannot be expressed as proportions of numbers. The easiest such "monster" to demonstrate is the ratio of the diagonal of a square to its side; now we say that...

√2 is irrational.

Zeno's Paradoxes

I, Zeno of Elea (c. 450 BC) was famous for my paradoxes, in which I challenged the foundations of our thinking about space, time and change.

With four paradoxes, Zeno tried to show that whether we conceive space as finitely or infinitely divisible, and whether we consider simple or relative motion, we get contradictions.

The best-known paradox concerns Achilles (the fastest runner) chasing a tortoise. In one jump, he halves the tortoise's lead; and then again; and then again …

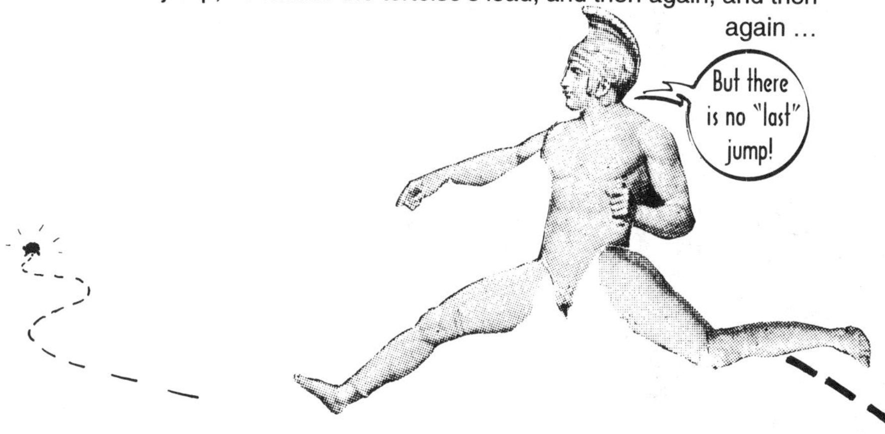
But there is no "last" jump!

Given this analysis, how do we describe his overtaking the tortoise?

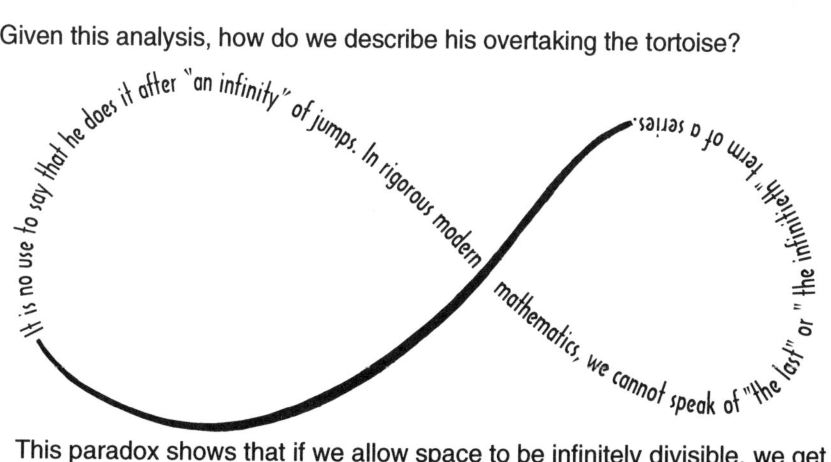
It is no use to say that he does it after "an infinity" of jumps. In rigorous modern mathematics, we cannot speak of "the last" or "the infinitieth" term of a series.

This paradox shows that if we allow space to be infinitely divisible, we get paradoxes in describing motion.

Zeno had three other paradoxes about motion, and others about change in general. Here is an example. Suppose we are given these instructions...

First, pour a cupful of wine into an empty barrel.
Then add one drop of water, and taste it to see if the mixture is still wine.
Repeat the operation until you come to the drop that changes it to water, then stop.
It's not wine anymore – it's become flavoured water!
If we don't know when we are on the boundary between two situations or things, how can we possibly say that they are different?

Of course, there is no such drop, but when the barrel is full we say:

However, we couldn't mark the transition point. Then, Zeno argues:

Philosophers have been chasing Zeno ever since he lived, but like Achilles they never quite catch their quarry. Perhaps he has something to tell us about our mathematical concepts. We like to believe they are clear, but perhaps they really are contradictory.

Euclid

I, Euclid (323-285 BC), am the father of demonstrative geometry.

His ideas had a huge impact upon Western mathematics, being the basis of our geometry until very recently. He systematized a tradition of proofs based on "constructions", using idealized instruments like a ruler and compasses (for making arcs of circles). With these, you could prove things about figures and their shapes, without using numerical examples. This was the big change in Greek mathematics – the idea of a proof that was general and, in its way, abstract.

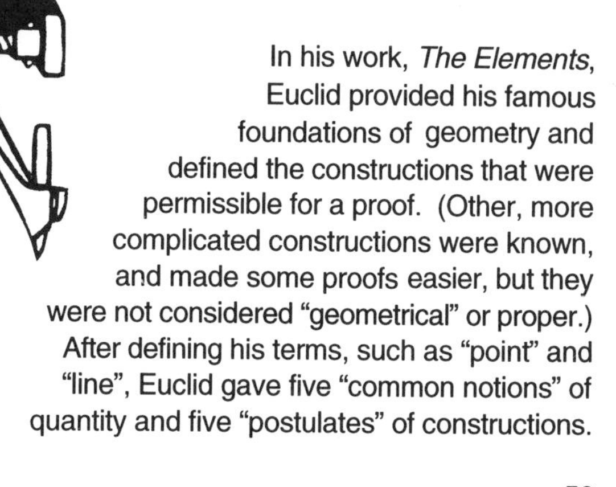

In his work, *The Elements*, Euclid provided his famous foundations of geometry and defined the constructions that were permissible for a proof. (Other, more complicated constructions were known, and made some proofs easier, but they were not considered "geometrical" or proper.) After defining his terms, such as "point" and "line", Euclid gave five "common notions" of quantity and five "postulates" of constructions.

The common notions:

1. Two things equal to a third are equal to each other $a=c$, $b=c$, $a=b$

2. Equals added to equals make equals = + = = =

3. Equals subtracted from equals make equals = − = = =

4. Two things which coincide are equal to each other

5. The whole is greater than the part

WHO_E
PART

The postulates:

Let it be granted that, in a plane

1. A line can be drawn between any two points

2. Any line can be extended in either direction indefinitely

3. A circle of any radius can be drawn about any centre

4. All right angles are equal

5. Two lines which cross a third line with interior angles whose sum is less than two right angles will meet

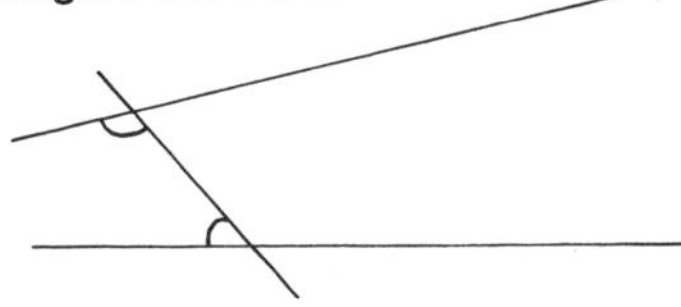

The first three define constructions, but the last two "postulates" are really theorems. The fifth postulate, also called "the parallel postulate", was a constant challenge to later mathematicians. Eventually, it proved to be the key to describing different sorts of geometries.

On this basis, Euclid went on to deduce all geometrical results known in his time, including the "Pythagoras's theorem". In spite of their obvious difficulties, his axioms were later regarded as self-evident truths, and the conclusions drawn from them likewise came to be treated as truths. Geometry was accepted as the great example of genuine knowledge that could be attained by human reason alone.

After Euclid, another very great mathematician was **Archimedes** (287-212 BC). He devised ways to measure the area of a number of curved figures, as well as the surface areas and volumes of a number of solids like spheres and cylinders. He worked out an approximate value of π...

...and I discovered the principle of displacement!

  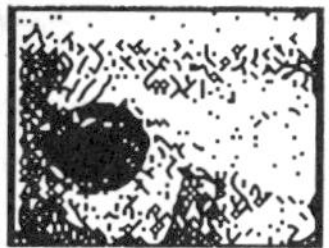   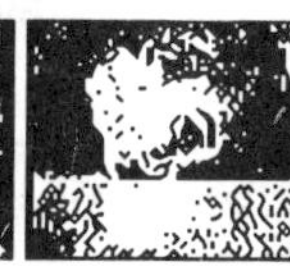

Chinese Mathematics

Forget the theory, as long as we can get a decent house out of these sums...

The Chinese never evolved the rigid style of proofs that we find in Euclid's *Elements*, because they were not really interested in formal logic. They were more concerned with the practical application of ideas and did not study mathematics for its own sake.

This did not prevent them from inventing their own proof for the sides of a right-angled triangle that was quite different from Pythagoras' theorem. And unlike the Greeks, they were not too bothered with surds (a number that cannot be expressed as a ratio of two whole numbers) or irrational numbers. To designate negative numbers, for example, the Chinese simply used red rods instead of black ones!

The Chinese practised algebra without the use of symbols, writing their ideas fully in words. They used a counting board for algebra as well as for other mathematical explorations. By the Sung Dynasty (960-1279), they had developed a notation that could handle equations as high as x^9. They could solve simultaneous linear equations (with two or more unknown quantities) and quadratic equations.

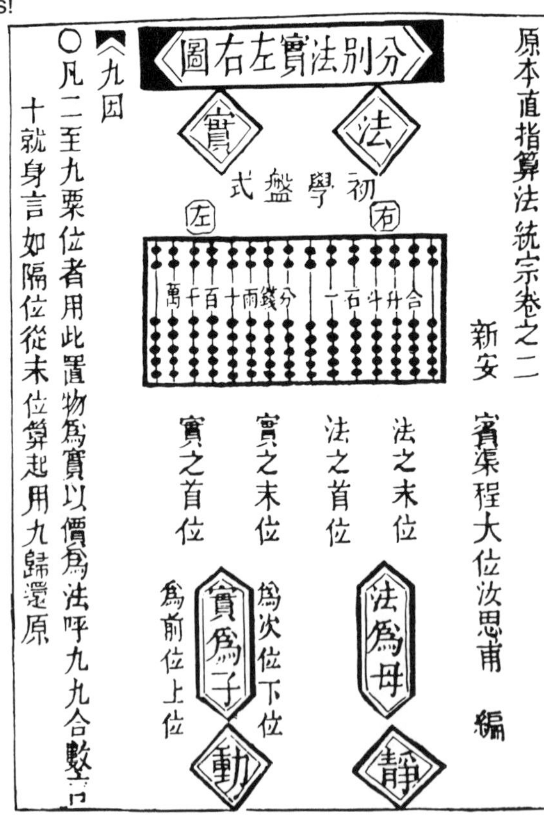
原本直指算法統宗卷之二
新安 賓渠程大位汝思甫 編
分別法實左右圖
法
實
初學盤式
右
左
法之末位
法之首位
實之末位
實之首位
法爲母
實爲子
爲次位下位
爲前位上位
靜
動
九因
凡二至九乘位者用此置物爲實以價爲法呼九九合數言十就身言如隔位從末位算起用九歸還原

4	9	2
3	5	7
8	1	6

They were also interested in "magic squares", their compartments filled up with numbers which all add up to the same total. This applies for horizontal rows, vertical rows and diagonal rows. They even devised three-dimensional cubes.

The Chinese were very keen to get an accurate value of π. **Liu Hui**, one of the earliest Chinese mathematicians, estimated π correct to four decimal places. His technique was to use the "method of exhaustion" where a polygon would be inserted inside a circle, and its number of sides increased until they became so short that it became possible to equate the polygon with the circle.

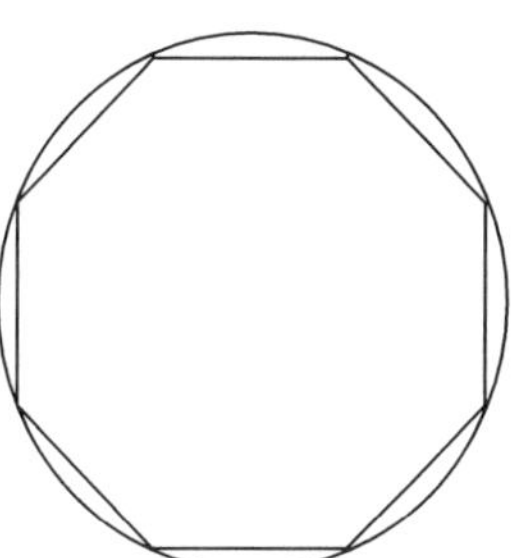

I also showed that the area of the circle is half the product of its circumference and its radius.

In the 5th century AD, the father and son team, **Tsu Ch'ung-Chih** and **Tsu Keng-Chih**, obtained a value for π of 3.1415926 and 3.1415927. This figure was not reached in the West until the 17th century.

The *Chiu Chang*

The *Chiu Chang* is the most famous book of Chinese mathematics. We do not know who wrote it, or the exact date it was written, but it is assumed to be from the late Chin or early Han dynasties (1st century AD). It covers the following topics:-

- land surveying (with rules for addition and subtraction of fractions), proportions (percentages)
- distributions by proportions (arithmetic and geometric progressions, rule of three)
- land mensuration (finding square and cube roots with a geometric basis)
- a reference text for engineers (volumes of three-dimensional objects)
- fair taxes (time to transport something from a to b, and distribution)
- a section on "too much and not enough" (puzzles on distribution and shortfalls)
- methods of tables (solution of simultaneous equations with two or three unknowns using a table), and finally, right-angled triangles (twenty-four problems on solving length of sides)

The range and depth of the *Chiu Chang* shows us the sophistication of Chinese mathematics by the beginning of the Christian era in the West.

Four Chinese Mathematicians

The last half of the 13th century and the early 14th century can be regarded as the peak of Chinese mathematics. Four of China's most famous mathematicians lived during this period –

I, Li Yeh who was a recluse...
I, Yang Hui who was a civil servant...
and I, Chu Shih Chieh who was a wandering teacher.
I, Chin Chiu Shao loved women and mathematics in equal measure and was also an accomplished swordsman...

There were more than thirty mathematical schools across China, and mathematics was a compulsory subject for national public service examinations.

Chin Chiu Shao is regarded as one of the greatest Chinese mathematicians ever; he worked for the military and the civil service. His book *Shu Chiu Chang* (Nine Sections of Mathematics) included some novel ideas, and introduced indeterminate analysis for the first time. (This is the study of problems whose solutions must be integers.)

Yang Hui and Chu Shih Chieh investigated permutations and combinations of expression, and came up with what we now call the Binomial Theorem. This involved multiplying two-term (binomial) expressions such as $(x + 1)$ and $(x + 3)$. This gives us $x^2 + 4x + 3 = 0$. The more expressions that are multiplied, the greater the number of terms in the solution, e.g.: $(x + 1)^3 = (x + 1)(x + 1)(x + 1) = x^3 + 3x^2 + 3x + 1$.

This led the two mathematicians to work on what we now call Pascal's triangle. They discovered that if one looks at the numbers in front of the x's, a pattern appears. Those for power one [i.e. $(x + 1)$] are 1, 1; those for power two [i.e. $(x + 1)^2$] are 1, 2, 1; those for power three [i.e. $(x + 1)^3$] are 1, 3, 3, 1; and so on. This is laid out in the form that Blaise Pascal designed in the 17th century.

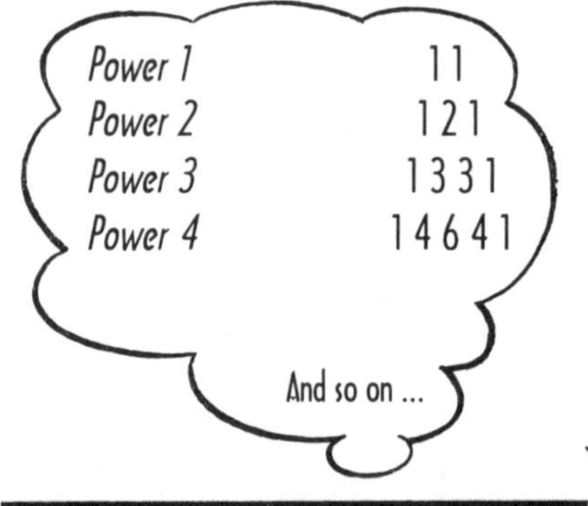
Power 1 11
Power 2 121
Power 3 1331
Power 4 14641
And so on ...

PASCAL

Pascal's triangle is used in the analysis of probabilities. The second row shows the number of permutations that could occur when two coins are tossed. There is one way to get two heads, two ways of getting a head and a tail, and one way of getting two tails.

The Chinese knew of this triangle five centuries before Pascal (1623-62).

It was first explained by the Sung mathematician **Chia Hsien** (c. 1100), and it could have appeared even earlier.

INDIAN
MATHEMATICS

Like the Chinese, the Indian mathematics relies on all varieties of proofs, including visual demonstrations that are not formulated with reference to any formal deductive system. Indian mathematics evolved from the framework developed by Indian logicians and linguists.

Mathematics in India developed in four distinct phases.

The Harappan Period from 2500 BC to about 1000 BC involved proto-mathematics for the use of bricks, etc.

This was followed by the Vedic Period, which lasted for about 1,000 years, and was concerned with ritual geometry. Jainism and Buddhism also begin to rise during this period.

The Classical Period then followed – this lasted until around 1000 AD. Mathematicians of this era were concerned with developing earlier concepts such as numbers, algorithms and algebra.

बाले मरालकुलमूल दलानिसप्त
तीरे विलास भरमंथरगाण्यपश्यम्
कुर्वचकेलि कलहं कलहंसयुग्मम्
शेषं जले वदमरालकुल प्रमाणम्

Poem from Indian mathematician Bhaskara (see facing page)...

The last great era for Indian mathematics was the Medieval Period of the Kerala School, which ended in the 1500s, in which earlier ideas were brilliantly developed. Quite why mathematics should have taken off in Kerala at this period is not known. However, it has been suggested that the Kerala School may have influenced European mathematics, as later "discoveries" in Europe had been anticipated by the Kerala mathematicians three centuries earlier.

Vedic Geometry

The Vedic Hindus were very fond of extremely large numbers, which formed part of their religious purview. For example, when discussing sacrifice, numbers such as 100,000 million are mentioned. There is a clear concept of numbers gradually rising in multiples of ten – the larger they were, the more interesting they became.

Altar geometry gives us an insight into the algebra of the Vedic Hindus. According to one system, the altar had the shape of an isosceles trapezium, and the sides had to be proportionally increased or decreased for various ceremonies. Further ceremonies required that some sides remained unchanged, while others had to be increased or decreased.

This provided the religious leaders with a mathematical problem, which required algebraic solutions. Rules were given for these operations, and questions regarding the number of bricks to be used in these alterations were also addressed. Deciding on how many bricks needed to be used, so that rifts in consecutive layers did not coincide, led to the use of simultaneous equations.

O girl! Out of a group of swans, 7/2 times the square root of the number are playing on the shore... the two remaining ones are playing with amorous fight, in the water. What is the total number of swans?
Mental arithmetic at a time like this, already...
Hint: Try the numbers N for which (N – 2)/ 7 is an integer!

Hindu mathematicians calculated the value of π correct to four decimal places.

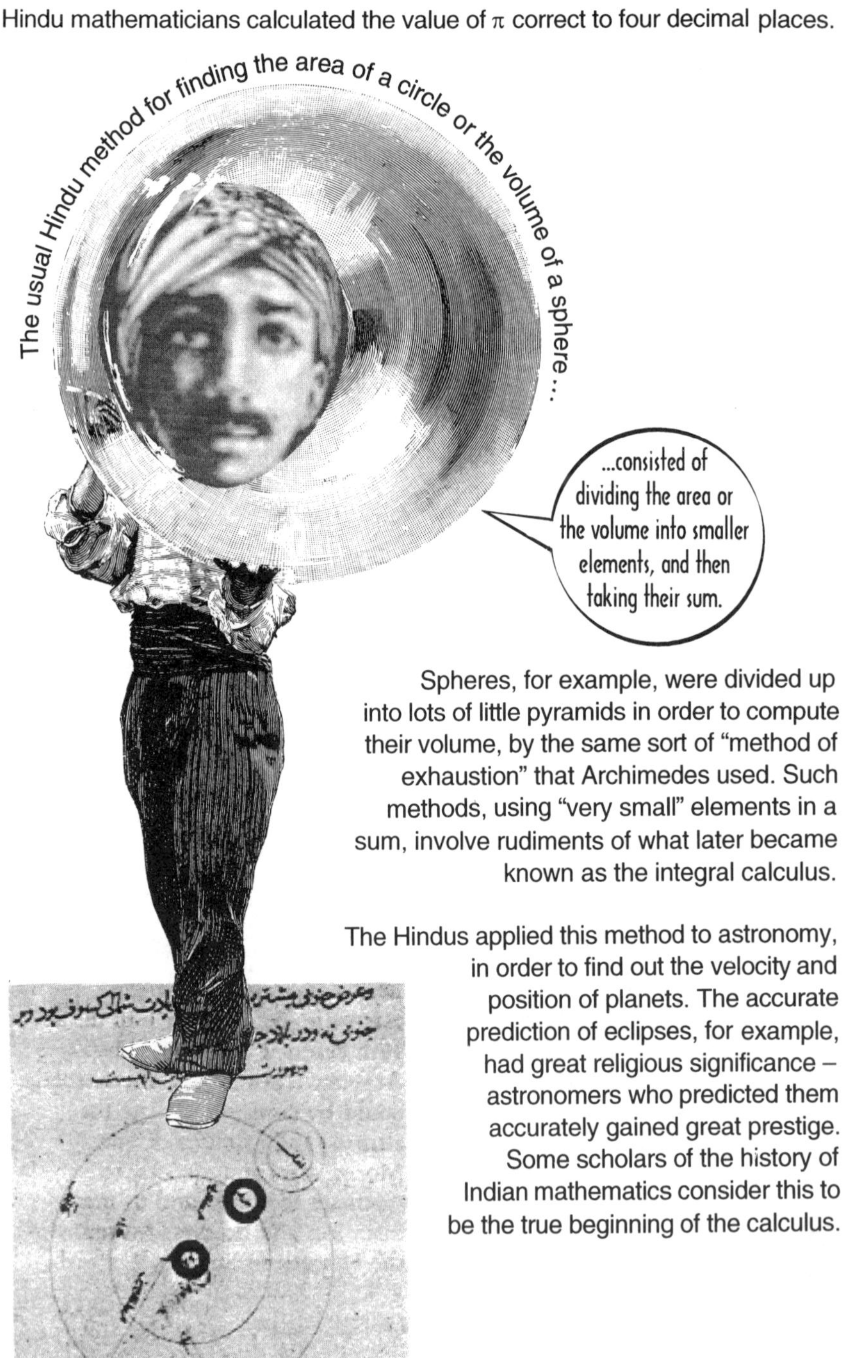
The usual Hindu method for finding the area of a circle or the volume of a sphere …
…consisted of dividing the area or the volume into smaller elements, and then taking their sum.

Spheres, for example, were divided up into lots of little pyramids in order to compute their volume, by the same sort of "method of exhaustion" that Archimedes used. Such methods, using "very small" elements in a sum, involve rudiments of what later became known as the integral calculus.

The Hindus applied this method to astronomy, in order to find out the velocity and position of planets. The accurate prediction of eclipses, for example, had great religious significance – astronomers who predicted them accurately gained great prestige. Some scholars of the history of Indian mathematics consider this to be the true beginning of the calculus.

Brahmagupta

Algebra later appeared as a separate branch of mathematics in the time of **Brahmagupta** (c. 598), one of the greatest of Indian mathematicians. He wrote a mathematical treatise in which he covered subjects such as square and cube roots, fractions, rules of three, five, seven, etc., and barter. During his time, equations were classified into groups we would recognize today: simple (*yavat-tavat*), quadratic (*varga*), cubic (*ghana*) and bi-quadratic (*varga-varga*). Brahmagupta was concerned with linear equations with unknowns, and quadratic equations. He had many commentators, who passed his ideas on through the years.

Like most Vedic Hindus, Brahmagupta loved irrational numbers such as $\sqrt{2}$, and gave their values up to a high degree of approximation.

Jain Numbers

Like the Vedic Hindus, Jains were also interested in extremely large numbers – and had a unique way of thinking about them. They suggested that there were three groups of numbers: numerable, innumerable and infinite. Each group was divided into three. The first group consisted of lowest, intermediate and highest numbers; the second consisted of nearly innumerable, truly innumerable and innumerably innumerable; the final group consisted of nearly infinite, truly infinite and infinitely infinite. European mathematics did not scale those heights until just a century ago, in the work of Cantor.

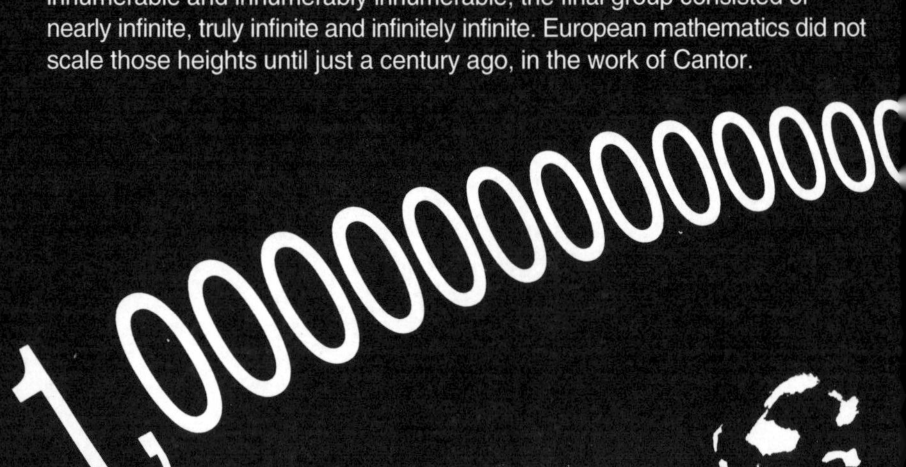

Mahaviracharya (c. 850), a Jain mathematician, used negative numbers in his work, and mentioned zero –

A number divided by zero remains unchanged.

It should be infinity.

Vedic and Jain Combinations

Both the Vedic Hindus and the Jains were fond of playing with combinations. One source of this interest could be the Vedic metres in poetry and their variations. Some metres have 6 syllables, some have more (e.g. 8, 9, 11 or 12). The challenge was to alter the long and short sounds within each syllable group, and to find the various combinations available. This search led to more games of permutations – e.g. the total number of perfumes that could be made from, say, 12 substances taken one, two, three or more at a time.

Sniff! This 8,11,9,3 smells terrible.
The result of this thought process was the meru-prastara, which is the same as Pascal's triangle.

Bhaskara II (c. 1114) used zero correctly in both his arithmetic and his algebra. For algebra, he employs the modern theory of using signs and letters to denote unknown quantities. He studied quite sophisticated problems in the theory of numbers, and his work is also credited with containing the "germ of the modern calculus".

Mathematical Verse

Mathematical ideas in India were often transmitted orally in verse form. Mathematical riddles in verse are common even today. A famous mathematical verse reads:

O beautiful maiden with beaming eyes, tell me, since you understand the method of inversion, what number multiplied by 3, then increased by three-quarters of the product, then divided by seven, then diminished by one third of the result, then multiplied by itself, then diminished by 52, whose square root is then extracted before 8 is added and then divided by 10, gives the final result of 2?
How long have I got?
It could be verse.
Oh no! Look on the next page! If you thought the poetry was bad, look at the maths!

The answer is 28. To obtain it, one works backwards through the riddle doing the inverse of the given operations. Thus we do, in order: X10, –8, $[\,]^2$, +52,etc.

Here's how the answer was reached:

$[(2)(10) - 8]^2 + 52$ which equals 196. Then,

$$\sqrt{196} = 14$$

Starting with this 14, we proceed:

$$\frac{(14)(3/2)(7)(4/7)}{3} = 28, \text{ the answer.}$$

Nowadays, we would start with x for the unknown answer, and write:

$$((\sqrt{\{[x.3.\ (7/4)(2/3)]^2 - 52\}} + 8))/3 = 2$$

Unscrambling that complicated expression is not so different from the old-fashioned way, but now we can keep an eye on the x until it stands alone equal to a number.

Ramanujan

The history of Indian mathematics is full of examples of intuitive mathematicians. **Srinivasa Ramanujan** (1887-1920), for example, was a complete academic failure but a brilliant mathematician. A humble accountant and a totally traditional man, Ramanujan relied on mysticism and metaphysics as much as abstract ideas for his mathematics. It was beyond anyone's comprehension how he arrived at his brilliant and profound (and occasionally wrong) results.

His patron in England, the mathematician **G.H. Hardy**, once visited him while he was in hospital.

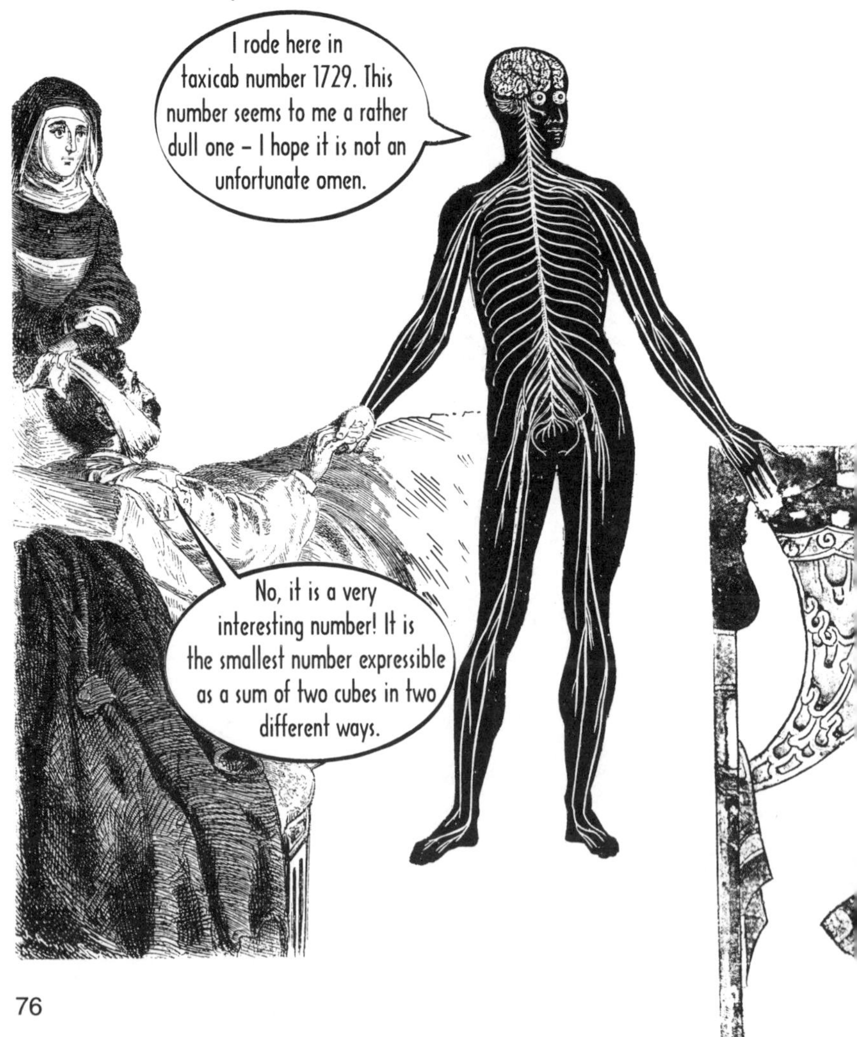
I rode here in taxicab number 1729. This number seems to me a rather dull one – I hope it is not an unfortunate omen.
No, it is a very interesting number! It is the smallest number expressible as a sum of two cubes in two different ways.

Islamic Mathematics

The Muslims unified the mathematical thought of earlier civilizations, fusing the algebraic and arithmetic traditions of Babylonia, India and China with the geometric traditions of Greece and the Hellenistic world. As a result, Muslim mathematicians were very confident at handling the basic arithmetical operations for both whole numbers and fractions, the use and interchangeability of decimal and sexagesimal numbers, the extraction of square roots and operations with irrational numbers, the extraction of cube roots, the elaboration of the binomial coefficients, and the extraction of the fourth and higher roots.

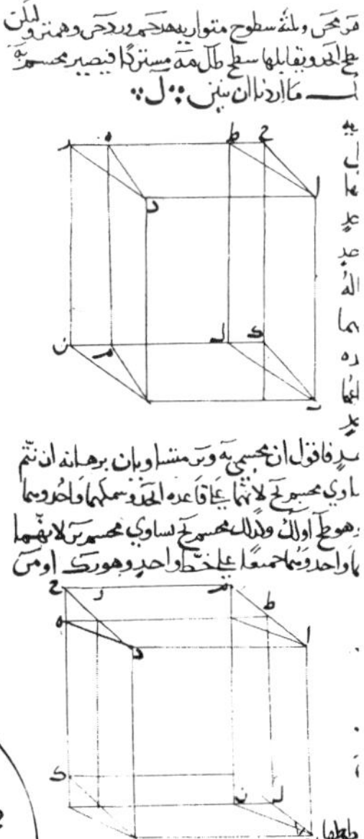

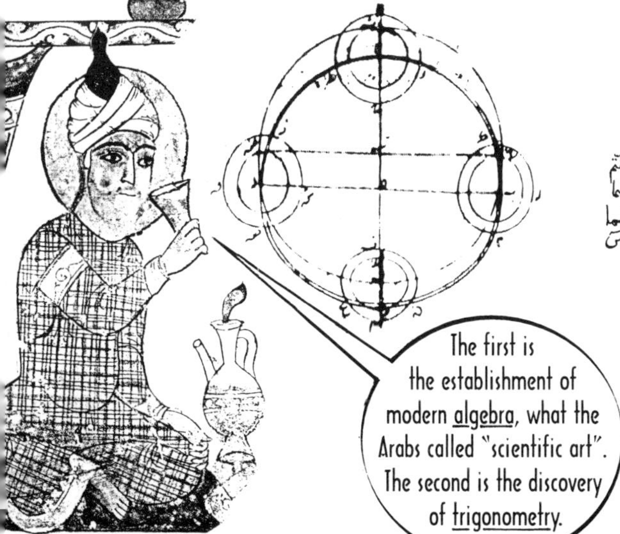

The Muslim mathematicians have two great achievements to their name.
The first is the establishment of modern algebra, what the Arabs called "scientific art". The second is the discovery of trigonometry.

Al-Khuwarazmi

Muhammad bin Musa al-Khuwarazmi (d. 847) was the founder of algebra as we know it today. The word "algebra" comes from the title of his book: *Kitab al-mukhasar fi hisab al-jabr wa'l muqabala* (*The Book of summary concerning calculating by transposition and reduction*). The word "algorithm" is derived from his name. Al-Khuwarazmi explained how it is possible to reduce any problem to one of six standard forms, using two processes, the first known as *al-jabr*, the second as *al-muqabala*.

Al-jabr was concerned with "transferring terms" to eliminate negative quantities (so that, for example, $x = 40 - 4x$ becomes $5x = 40$).

Al-muqabala was the next process, that of "balancing" the positive quantities that remain (thus, if we have $50 + x^2 = 29 + 10x$, *al-muqabala* reduces it to $x^2 + 21 = 10x$).

كتاب المختصر في حساب الجبر والمقابلة
ابو جعفر محمد بن موسى الخوارزمي
بسم الله الرحمن الرحيم
هذا كتاب وضعه محمد بن

In his book, al-Khuwarazmi used no symbols as we do now – these came later – and he expressed his mathematics in words. Using words, he described the solutions to quadratic equations and developed the now standard formula:

$$ax^2 + bx + c = 0$$

which has the solution:

$$x = [1/2a][-b \pm \sqrt{(b^2 - 4ac)}]$$

We met this before on page 45

Development of Algebra

Muslim mathematicians set out deliberately "to operate on unknowns with the aid of all the arithmetical tools available, just as the arithmetician operates on knowns".
He also had immense ability for handling negative numbers, which he treated as separate entities.

For us, the purpose of algebra was twofold: the systematic application of the operations of elementary arithmetic to algebraic expressions; and the study of algebraic expressions independently of what they represented, so as to be able to apply to them the general operations that had been applied to numbers.

Al-Samaw'al (d. 1175)

Al-Samaw'al was the first to write algebraic results in symbolic forms.

In my book al-Fakhri, I studied the different "powers of the unknown".
I also applied arithmetical operations to algebraic expressions, and produced one of the first statements of the algebra of "polynomials".
Al-Karaji
(d. 1000)
Al-Karaji's work was used by his successors to propose rules for the divisibility of polynomials, one by another; and also rules for the possibility of extracting a "square root" of a polynomial.
Out of this came "combinatorial analysis", later applied to the analysis of games of chance, calculating the probabilities of the arrangements of dice or cards.
A formula for "binomial expansion" was also enunciated and derived.
Its table of coefficients, "Pascal's triangle" (already discovered by Chinese mathematicians), was exhibited.

Omar al-Khayyam (d. 1123) also discussed finding the roots of the fourth, fifth, sixth and higher powers by a method he had discovered that did not involve using geometry, but an equivalent of Pascal's triangle. His discovery was contemporaneous with that in China.
I also wrote poetry as a sideline!
I produced a book on algebra written in verse, and made algebraic symbols more widely known in the West.
Abu'l Hasan al-Qalasadi (d. 1486)
Besides calculating π correct to sixteen decimal places, **al-Kashi** (d. 1429) introduced methodical ways of dealing with decimal fractions.

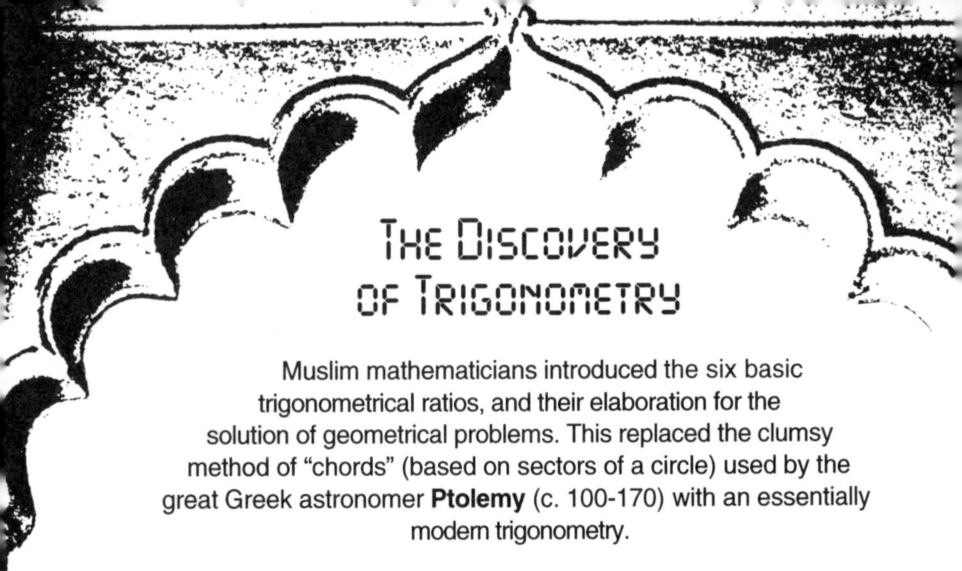

The Discovery of Trigonometry

Muslim mathematicians introduced the six basic trigonometrical ratios, and their elaboration for the solution of geometrical problems. This replaced the clumsy method of "chords" (based on sectors of a circle) used by the great Greek astronomer **Ptolemy** (c. 100-170) with an essentially modern trigonometry.

These "functions" are defined in terms of the sides of a right-angled triangle. We call them **O** for Opposite a particular angle, **A** for Adjacent to it, and **H** for Hypotenuse, the long side. Then **sine** = O/H, **cosine** = A/H, and **tangent** = O/A. An incredible world of relationships comes out of those three simple definitions. Trigonometry was a development of the greatest importance in the progress of mathematics, astronomy and practical arts such as surveying and fortification.

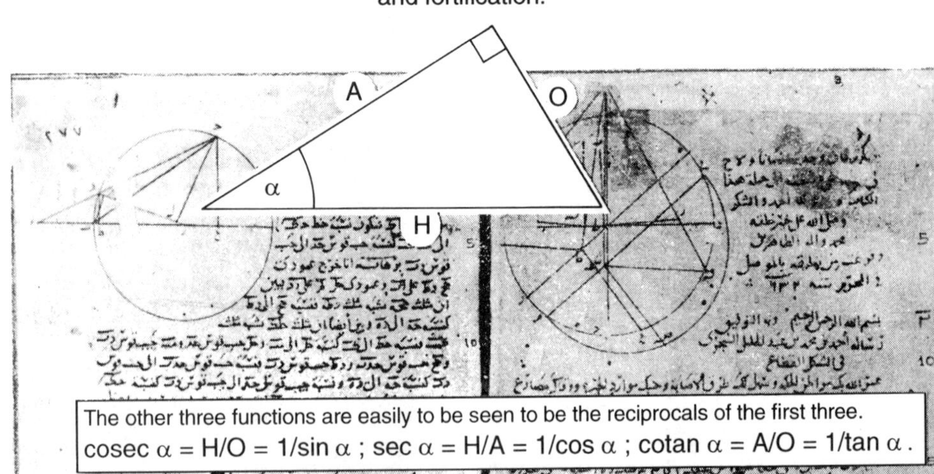

A
O
H
α

The other three functions are easily to be seen to be the reciprocals of the first three.
cosec α = H/O = 1/sin α ; sec α = H/A = 1/cos α ; cotan α = A/O = 1/tan α .

Al-Battani

Al-Battani (d. 929) produced a number of trigonometrical relationships. These included:

$\tan a = \sin a / \cos a$

$\sec a = \sqrt{1 + \tan^2 a}$

He also solved the equation $\sin x = a \cos x$, discovering the formula:

$\sin x = a / \sqrt{1 + a^2}$

I also used the idea of the tangent or "shadow", first introduced by al-Marwazi (c. 900), to develop equations for calculating tangents and cotangents, and compiled a table of cotangents.

Abu Wafa

Abu Wafa (d. 998) established the relations:

$$\sin(a+b) = \sin a \cos b + \cos a \sin b$$

$$\cos 2a = 1 - 2\sin^2 a$$

$$\sin 2a = 2 \sin a \cos a$$

and discovered the sine formula for spherical geometry:

$$\frac{\sin A}{\sin a} = \frac{\sin B}{\sin b} = \frac{\sin C}{\sin c}$$

My constructions were so eminently serviceable that they were widely circulated in Europe during the Renaissance. I also prepared new trigonometric tables, and developed ways of solving some problems of spherical triangles.

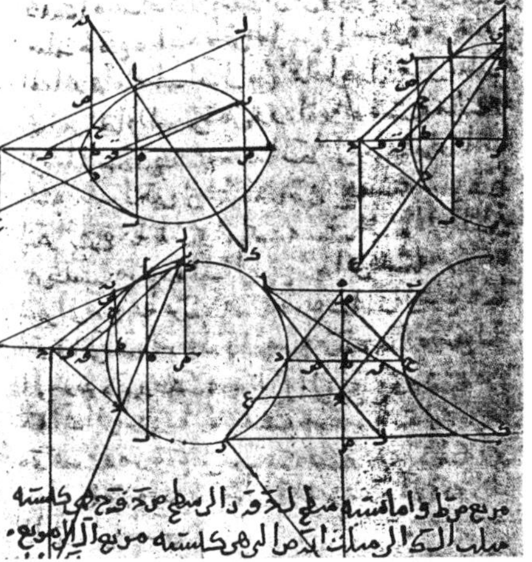

A,B,C are the lengths (in degrees) of the great-circles making up a triangle on the surface of the sphere and a,b,c are the angles opposite to them. Great-circles on a sphere are made by planes which pass through the centre of a sphere. (Nowadays transcontinental aircraft fly along great circles, as these define the shortest path between two points.)

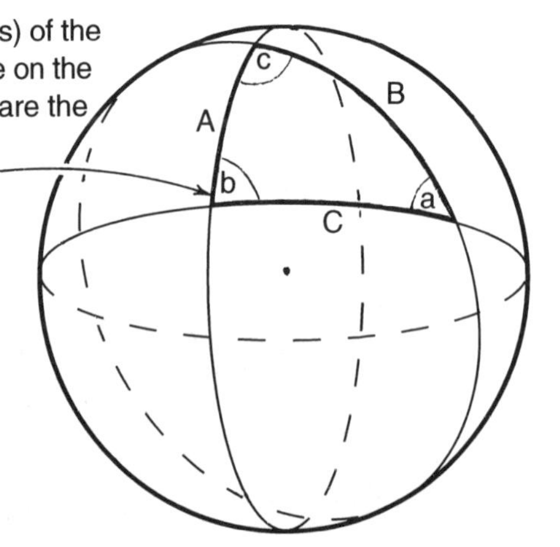
c
B
A
b
a
C

Ibn Yunus and Thabit ibn Qurra

Ibn Yunus (d. 1009) demonstrated the formula:

$$\cos a \cos b = 1/2 [\cos (a + b) = \cos (a - b)]$$

Although this deals with trigonometric functions, it enables a product to be evaluated as a sum. In the days when calculation of products of numbers with many digits was tedious, this was a great labour-saving device. Later, it gave a stimulus to the development of logarithms, which performed the same function more directly. It also led to the fundamental formula of spherical trigonometry in use today, the cosine formula:

$$\cos a = \cos b \cos c + \sin b \sin c \cos A$$

(where A is the arc of a great-circle, a is the angle opposite)

Thabit ibn Qurra (d. 901) wrote on the theory of numbers, and extended their use to describe the ratios between geometrical quantities – a step the Greeks never took.

He also discussed the question of where, if anywhere, parallel lines can meet.

Al-Tusi

The most eminent scholar in the field of trigonometry, both plane and spherical, was **Nasir al-Din al-Tusi** (d. 1274).
His comprehensive treatment of the resolution of spherical triangles is one of the landmark studies in the development of mathematics.
He formulated the "Tusi couple" which showed how a linear back-and-forth motion could be represented as the composition of two circular motions.
This device enabled **Nicolas Copernicus** (1473-1543) to represent all the irregular motions of the planets as compounded circular motions, and that made it easier for him to construct an astronomical system with the sun, and not the earth, at the centre.

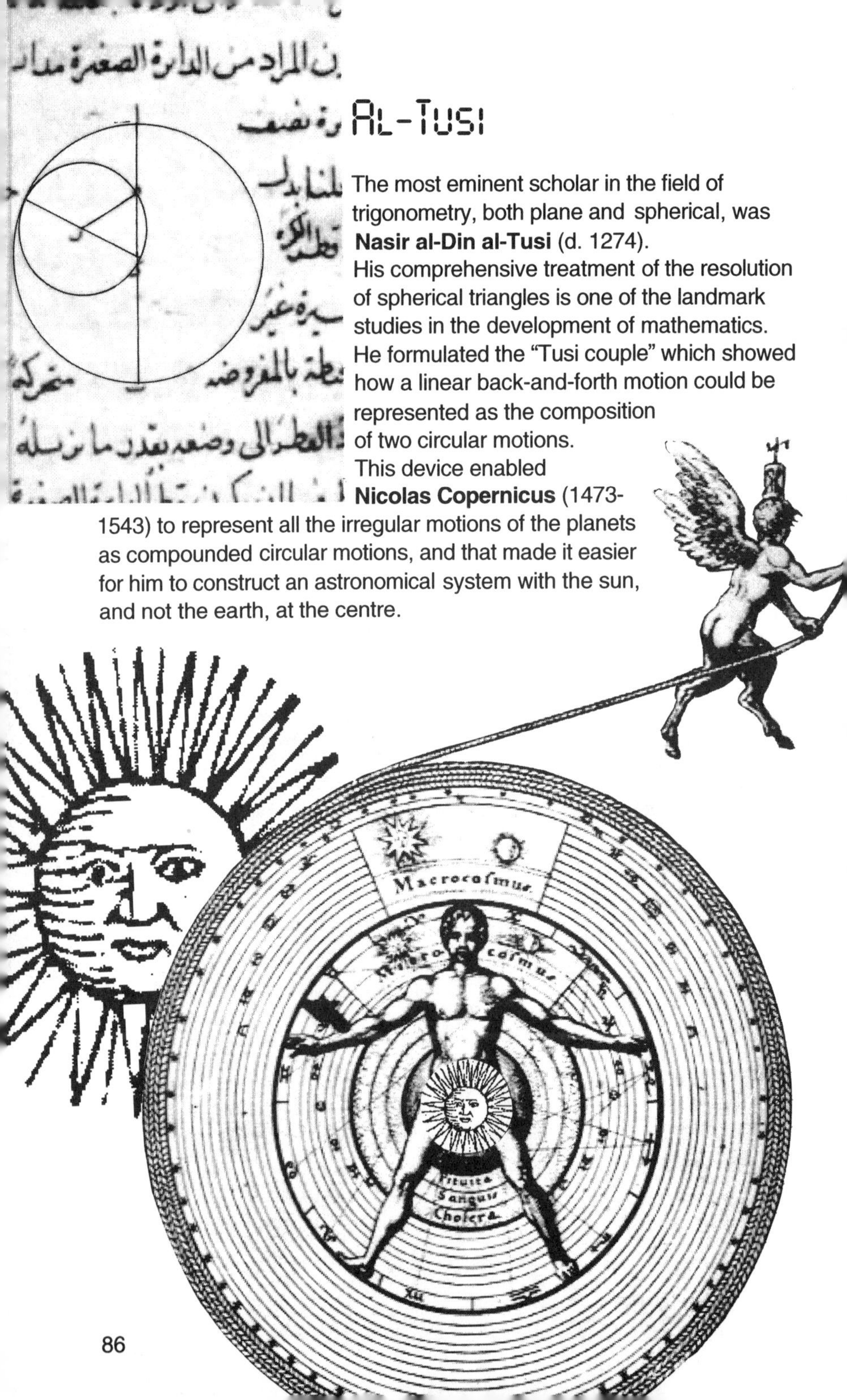
Macrocosmus.
Micro cosmus
Pituita
Sanguis
Cholera

Solutions of Problems Involving Integers

For many centuries, problems with integer solutions were very popular. After all, these were the "numbers" that people understand. An example is the "inheritance" problem:

Four sons inherit (respectively) a third, a fourth, a fifth and a sixth of their father's estate ...
...leaving five pieces of silver as the residue.
How big is the estate?
Problems like this are part of the stock of "riddles" used to challenge the wits (and bravery) of opponents such as we, and are normally solved by trial-and-error.
Actually, I know the answer's 100.

A systematic approach to such problems was first achieved by **Diophantus** (c. 275). Muslim mathematicians were active in the theoretical development of this work. A natural starting point were the "Pythagorean numbers" like 3, 4, 5, which form the sides of a right-angled triangle. Then the relationship was generalized, and in the 10th century, Muslim mathematicians asked: can the equation $x^n + y^n = z^n$ be solved in whole numbers? Like Fermat several centuries later (after whom the problem is named), several mathematicians thought they had a proof of the impossibility of a solution. Their successors discovered their errors, and now we know that it is a very difficult problem indeed!

Emergence of European Mathematics

European mathematics relied on the contributions of all other civilizations for its development. Throughout the Middle Ages, Europe was significantly inferior to the civilizations further East in terms of technology, science and culture. It gradually caught up, first through cultural contacts during the Crusades, and then through dialogues among scholars in Spain and Italy.

Arabic language materials (either translations from the Greek or original work) would be translated by teams, sometimes involving Jewish intermediaries.

I hope you enjoy the algebra as much as the alcohol.
Coo, thanks.

Scientific names beginning with "al-", like algebra and alcohol, are a reminder of that process. In the Renaissance of the 15th century, the Pythagorean tradition of aesthetic and mystical mathematics was rediscovered.

Then, in the 16th century "age of expansion", European mathematics took off.
Exploration, conquest and religious war were the great themes of that age.

Mathematics was needed for navigation overseas, and it was employed for defence (designing fortifications) and attack (artillery tables) at home. Fields like trigonometry were vital for success in such ventures. They progressed both in practice (better tables) and in theory.

Also, there was the steady development of commerce, calling for improved reckoning methods. Earlier, the Church had denounced the "Arabic" numerals, and double-entry book-keeping was seen as a magical art (not without justification, it must be admitted). But now they were too important to be neglected.

Progress in European theoretical mathematics was accompanied by a series of crises and paradoxes. Negative and irrational numbers, which hardly bothered Chinese, Indian and Islamic mathematicians, became highly problematic for European mathematicians, even while they used them with great success. Eventually, the paradoxes gave rise to new fields of mathematics ...

...which in the 20th century became the ultimate in paradox themselves.

René Descartes

It is significant that the greatest European innovator in mathematics, the Frenchman **René Descartes** (1596-1650), was also a philosopher. In his personal quest for certainty, he turned away from humanistic literary learning to pursue mathematics. But at first he was disappointed.

I found algebra to be obscure and confused, and geometry too restricted...
... so I set out to combine their strengths and remedy their weaknesses.

Why did Descartes refer so disparagingly to the algebra that he was determined to improve? Well, algebra was only partly formalized during the 16th century. Some of the terms had been given abbreviated names, which were neither clear descriptions nor tokens to be manipulated. But for the mathematicians of the time, there were worse difficulties. They found themselves describing things that were nonsensical, or worse!

We have already mentioned "imaginary" numbers, the roots of algebraic equations like $x^2 + 1 = 0$. What sort of numbers are these? We can't enumerate objects with them. What sorts of physical objects can there be, whose measurement, when squared, gives a negative quantity? It is all very well to manipulate them by the rules, but there is no security against writing nonsense.

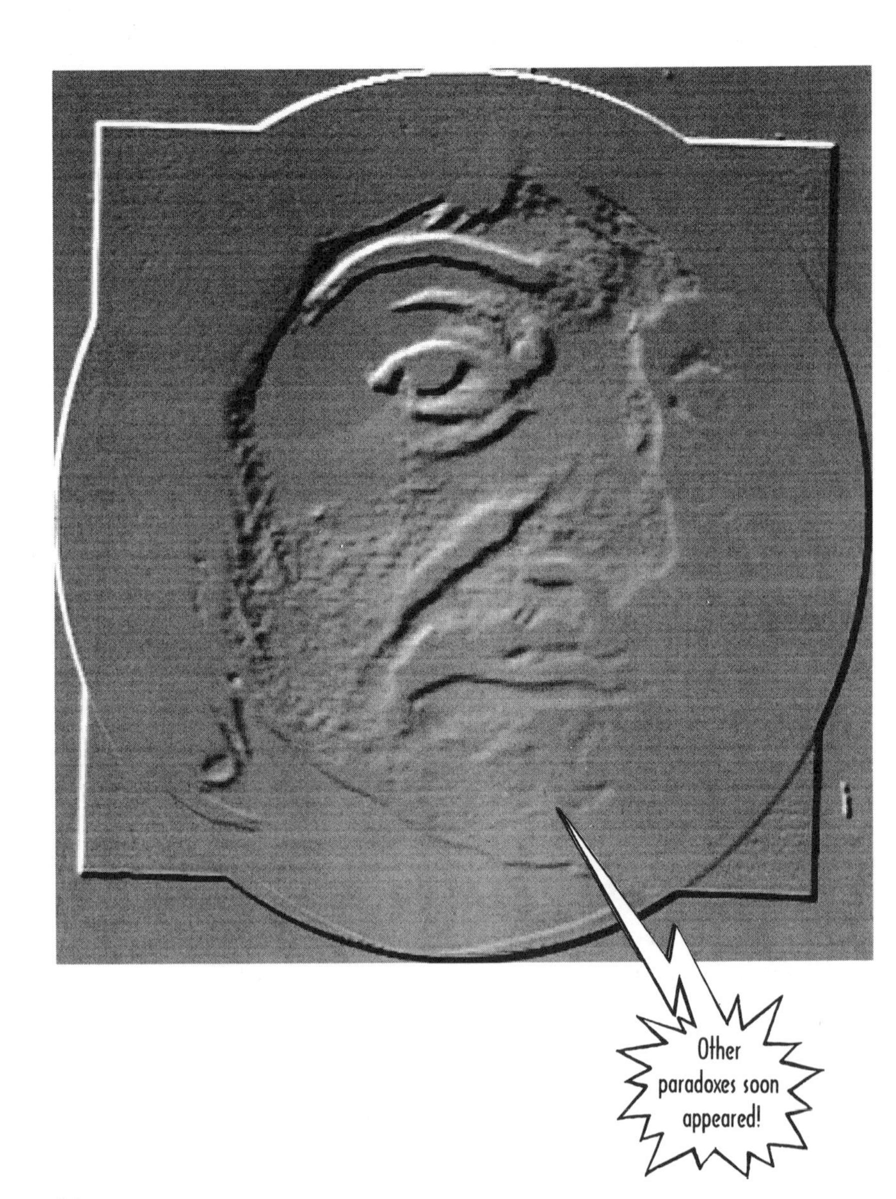
Other paradoxes soon appeared!

Analytic Geometry

Out of Descartes' efforts came "analytic" or "coordinate" geometry.

Analytic geometry is based on the idea that a point in space...

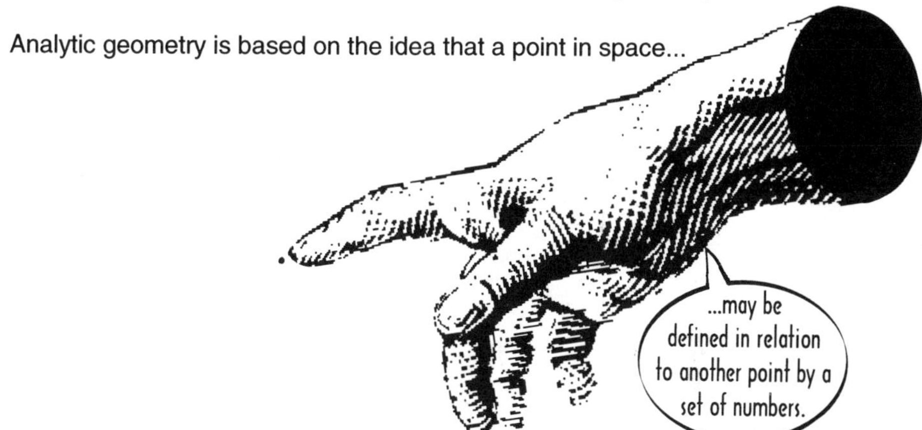
...may be defined in relation to another point by a set of numbers.

In plane geometry, there are two axes at right angles that we normally call x-axis and y-axis. The position of any point in the plane of axis can be given by its coordinates (x,y), which give its distance in the x and y directions from the origin – the point of intersection of the two axes.

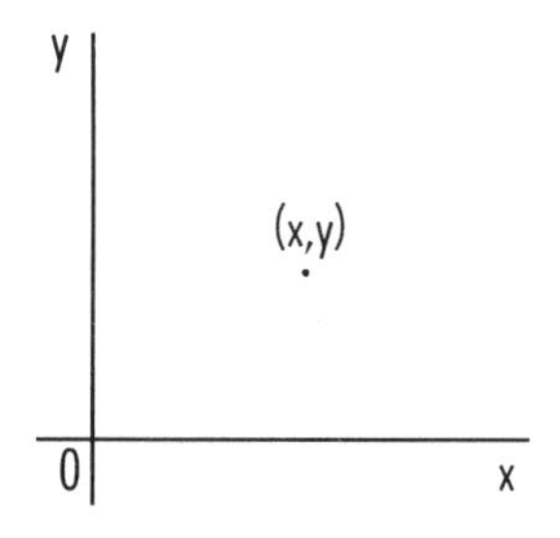
y
(x,y)
0
x

In three dimensions, we have three axes at mutual right angles: x-, y- and z-axes.

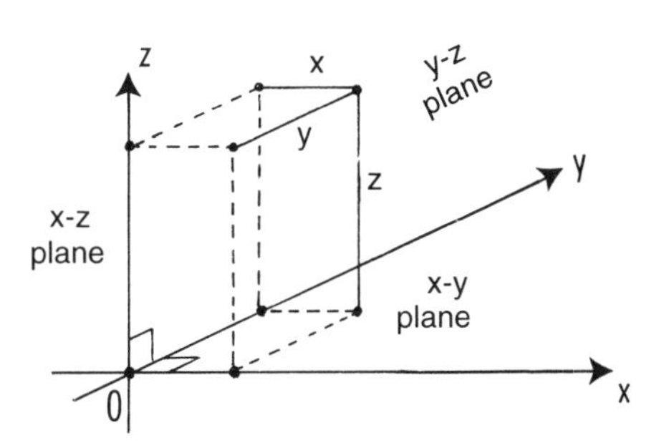
z
x
y-z plane
y
z
y
x-z plane
x-y plane
0
x

On the x- and y-axes, we can plot a graph, point by point.
Moreover, we can establish a relationship between the coordinates of every point with an equation.

The simplest form of a graph is a straight line, which is described by a linear equation of the form $y = ax + b$, where a and b are constants.

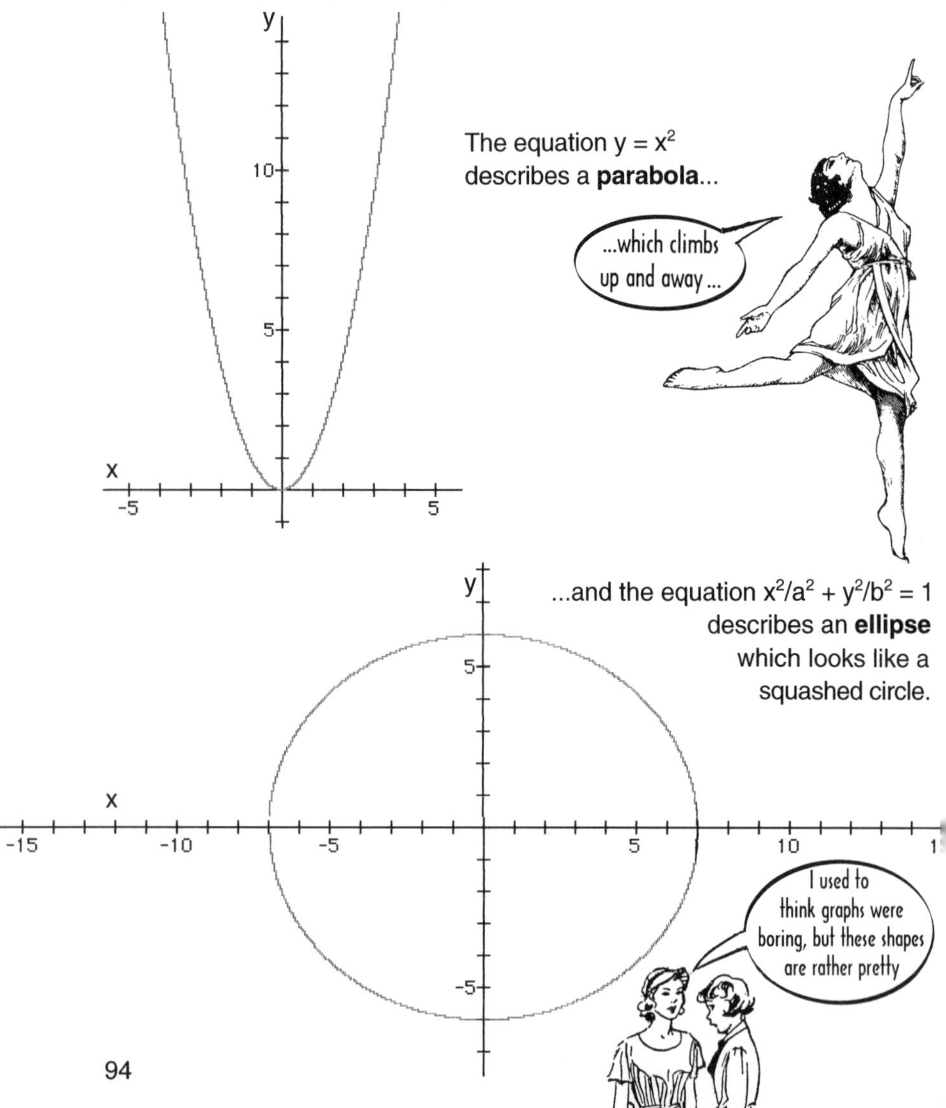
y
10
5
x
-5
5
The equation $y = x^2$ describes a **parabola**...
...which climbs up and away ...
y
...and the equation $x^2/a^2 + y^2/b^2 = 1$ describes an **ellipse** which looks like a squashed circle.
5
x
-15
-10
-5
5
10
-5
I used to think graphs were boring, but these shapes are rather pretty

The third curve in this family ...
...called the "conic sections"...

...is the **hyperbola**, with the equation $x^2/a^2 - y^2/b^2 = 1$. The minus sign makes all the difference, as this curve has two branches that climb away to infinity.

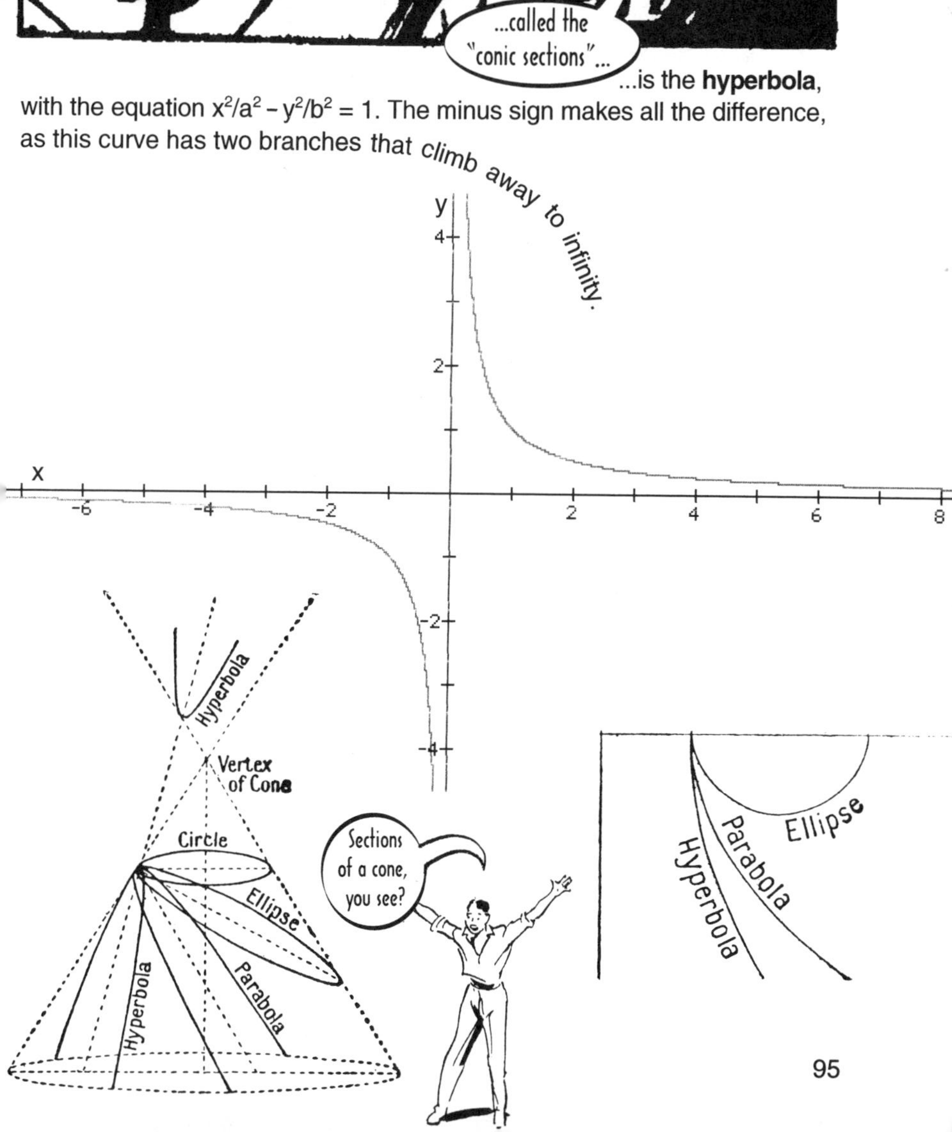
y
4
2
-2
-4
x
-6
-4
-2
2
4
6
8
Hyperbola
Vertex of Cone
Circle
Ellipse
Parabola
Hyperbola
Sections of a cone, you see?
Ellipse
Parabola
Hyperbola

Functions

The term "**function**" conveys a relationship, or dependence, of one variable term on another or others. We say, for example, that y is a function of x, or that z is a function of x and y. (Following Descartes, we use the letters at the end of the alphabet for variables and those at the beginning, such as a,b,c, for constants.)

X
Z
In analytical geometry and calculus we make use of functions described through certain symbols.

Thus, if the rule defining the function is:
square the number,
add twice the number,
subtract three, we write:
$f(x) = x^2 + 2x - 3$.

In analytical geometry, such a function of one variable may be plotted by setting x along one axis and f(x), the function of x, along the other. This function is a a parabola which crosses the x-axis at the points x = -3 and x = +1, and which has its lowest point at x = -1, y = -4.

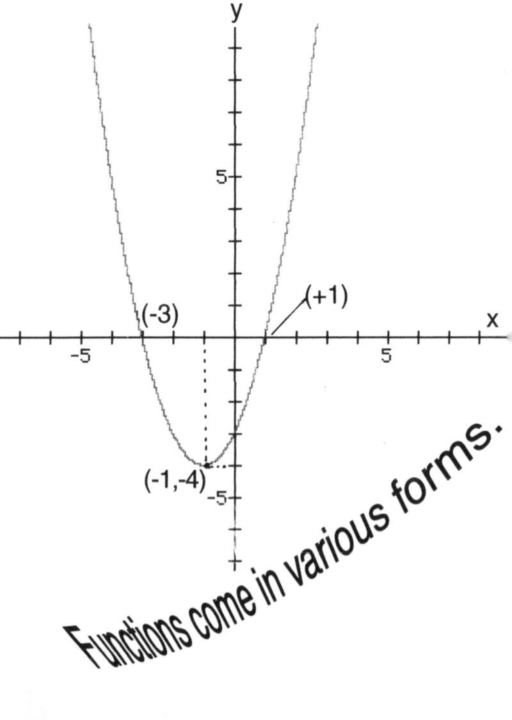
y
x
5
-10
-5
5
-5
(-3)
(+1)
(-1,-4)
Functions come in various forms.

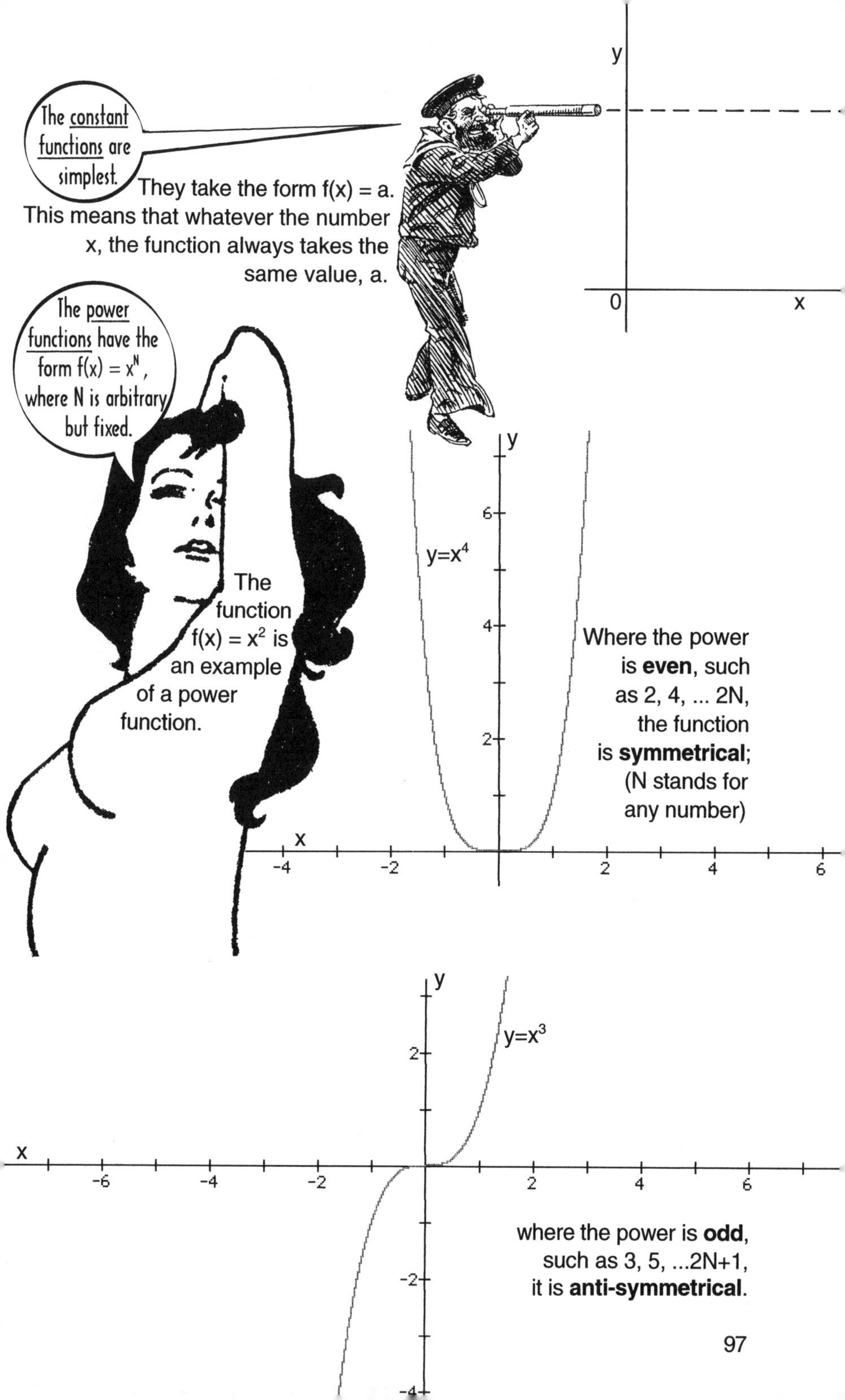
The constant functions are simplest.
They take the form f(x) = a. This means that whatever the number x, the function always takes the same value, a.
y
0
x
The power functions have the form f(x) = x^N, where N is arbitrary but fixed.
The function f(x) = x² is an example of a power function.
y
6
y=x⁴
4
2
x
-4
-2
2
4
6
Where the power is even, such as 2, 4, ... 2N, the function is symmetrical; (N stands for any number)
y
2
y=x³
x
-6
-4
-2
2
4
6
-2
-4
where the power is odd, such as 3, 5, ...2N+1, it is anti-symmetrical.

The **root functions** represent the "inverse" of the power functions; thus we have $f(x) = x^{1/2} = \sqrt{x}$, as the inverse of $f(x) = x^2$.

The **polynomial functions** have fixed numbers, a, b, c, d, etc., and a variable, x, that varies in its power. So a polynomial function may have the form $f(x) = ax^3 + bx^2 + cx + d$.

y=√x
x
y
2
-2
-6
-4
-2
2
4
6
Beyond these lies the tricky territory of the "transcendental" functions ...
... which transcend the realm of algebraic operations.

The **trigonometric functions** use the trigonometric ratios, such as sines and cosines. One such is $f(x) = \sin x$.

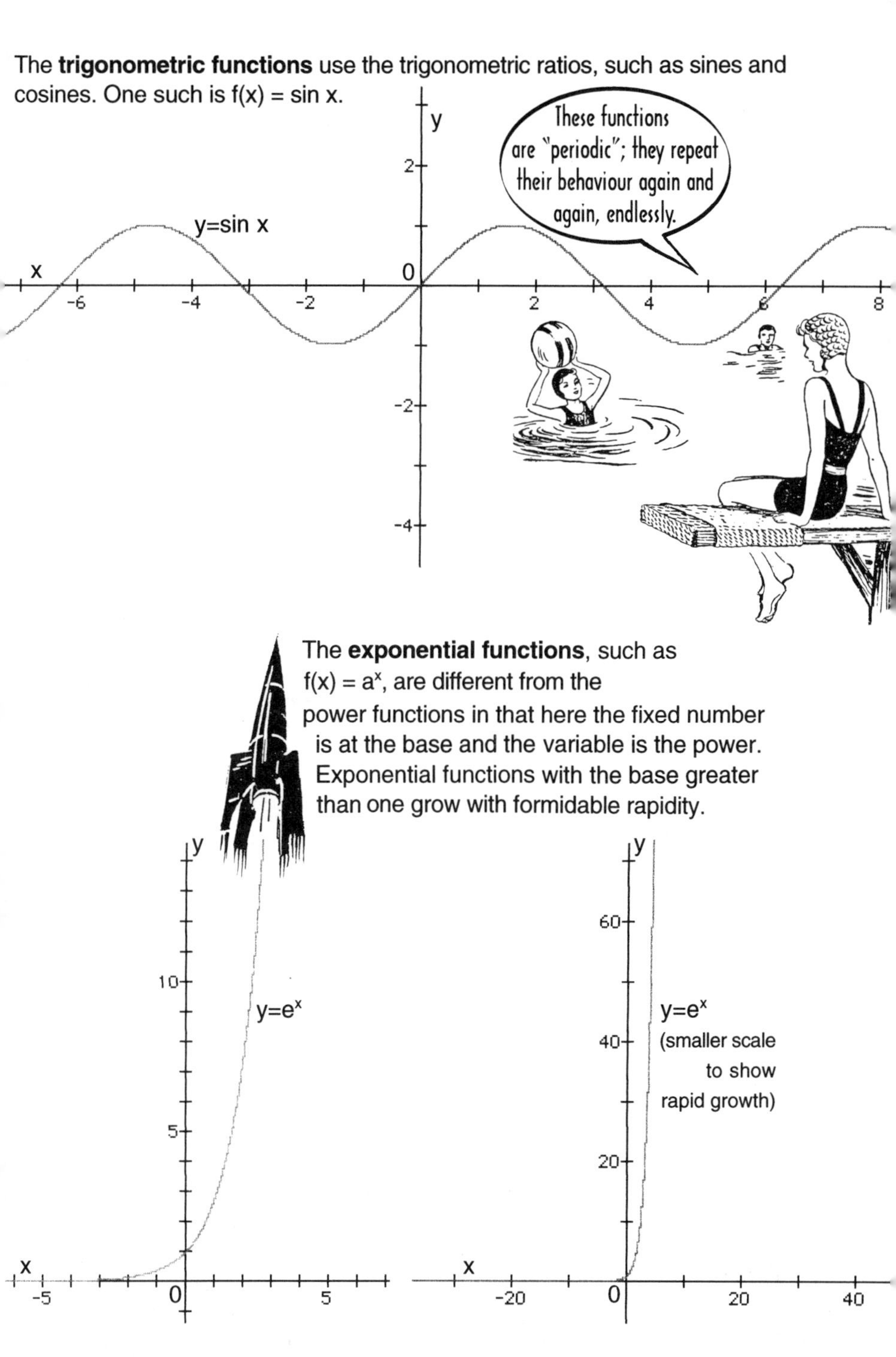
These functions are "periodic"; they repeat their behaviour again and again, endlessly.
y=sin x
y
x
2
0
-2
-4
-6
4
6
8
y=e^x
10
5
0
-5
y=e^x
(smaller scale to show rapid growth)
60
40
20
-20

The **exponential functions**, such as $f(x) = a^x$, are different from the power functions in that here the fixed number is at the base and the variable is the power. Exponential functions with the base greater than one grow with formidable rapidity.

The **logarithmic functions** are the inverses of the exponential functions. They are written $f(x) = Log_a(x)$; the number a is called the **base** of the logarithm. These functions grow very slowly, for example: $Log_a(10x) = Log_a(x) + Log_a(10)$.

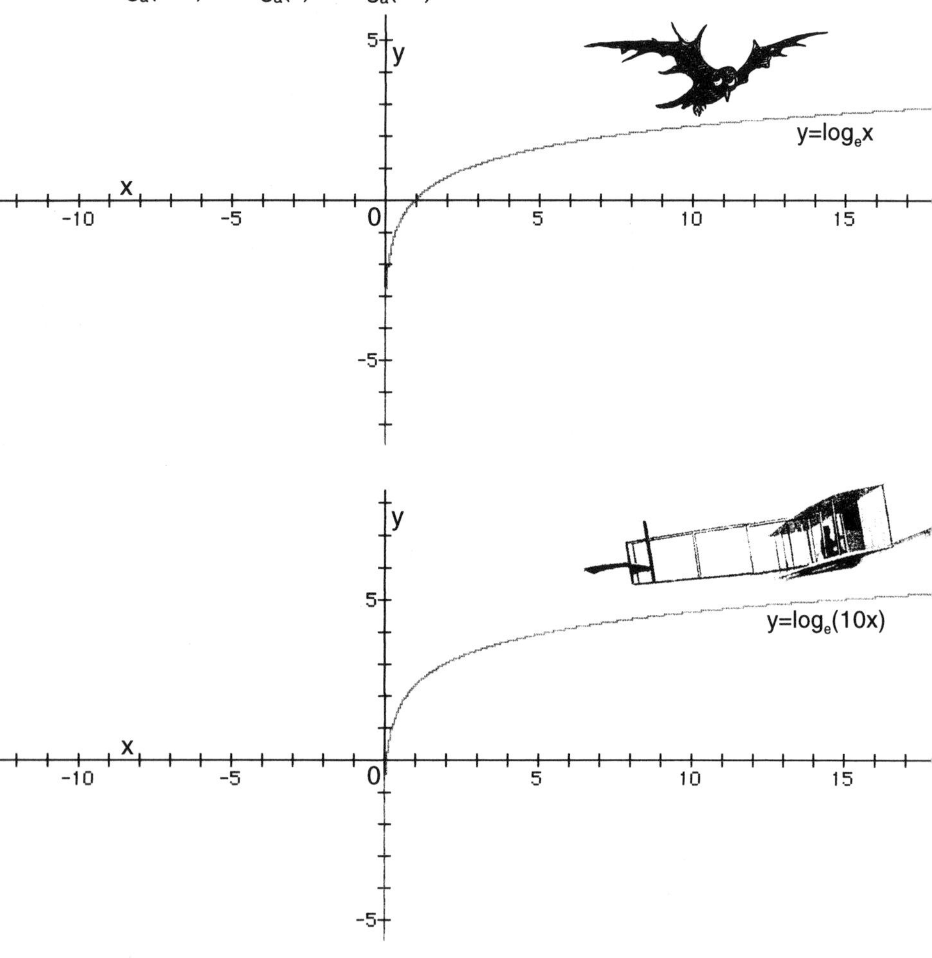

y
x
5
-5
-10
-5
0
5
10
15
y=log$_e$x
y
x
5
-5
-10
-5
0
5
10
15
y=log$_e$(10x)

The logarithms that we use in tables are taken to the base ten. In computers, which run on binary arithmetic (number 0,1) the appropriate base is two. But in theoretical mathematics the preferred base is the number e = 2.71828.... This is the "mother of all bases", representing the exponential function $f(x) = e^x$, which is the one whose rate of growth is exactly equal to its size.

Functions are the main analytical tools of the calculus.

The Calculus

Descartes' work was the culmination of a process of liberating algebra from words, rather as Greek geometry had liberated constructions from numbers. Once he had provided a formalism for describing algebraic relations, progress was swift. Within forty years of the publication of Descartes' algebraic geometry, the German philosopher and mathematician **Gottfried Wilhelm von Leibniz** (1646-1716) had created an algebra of the infinite. This is what we call "the calculus", a powerful tool for analysing growth and change.

The position of the "flowing" body: x

The speed, or "fluxion": $\dot{x}$.

Newton

The variable: x

The function: $f(x)$

The curve $y = f(x)$

The slope-of-tangent = derivative: $f'(x) = dy/dx$.

The area under the curve between points $x = a$ and $x = b$:

$\int_a^b f(x)dx$.

Leibniz

Sir Isaac Newton (1642-1727) had made an equivalent discovery somewhat earlier, but he merely extended Descartes' notation rather than going beyond it, so it is the Leibnizian form of calculus that predominates today. Thus, it was two philosophers, Descartes and Leibniz, who created the notations and ideas that have shaped mathematics ever since.

The secret of the calculus lay in unifying two sorts of problems that had previously seemed totally unrelated. We now call them differentiation and integration.

Differentiation

The calculus can be seen as an extension of analytic geometry, much of whose terminology it shares.
It deals with continuously varying quantities.

The process of finding how quickly a quantity changes is called **differentiation**. When we differentiate a function, we obtain its *rate of change*.

Think of a vehicle moving on a road. Its position is changing continuously along the road. At any given time, t, its position, x, is represented by the continuous function x(t).

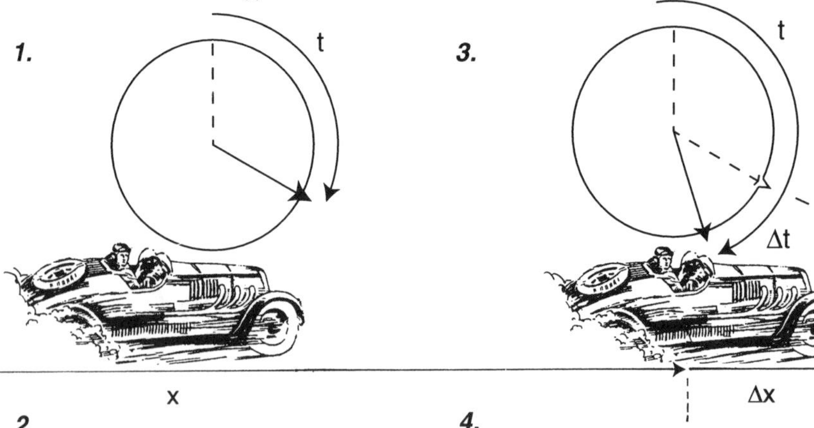
1.
t
3.
t
Δt
x
Δx

2.
The vehicle continues to move, and after an increment of time, let us call it Δt, it has moved to a new position, let us call it $x + \Delta x$.

4.
The vehicle arrives at its new position at a new time, which is the sum of the original time, t, and the amount of time it has travelled to get to the new position, in other words, $t + \Delta t$.

What is the average speed, or more technically, "velocity", of our vehicle? It is given by the distance travelled, Δx, divided by the time taken to travel it, Δt. Or:

$$\Delta x / \Delta t = f(t + \Delta t) - f(t) / \Delta t$$

So, suppose that we want to define the velocity of a moving body, say a car, at the instant t, or the rate of change of x at the time t. We can try to do this by making the increment Δt as small as possible, so small in fact that it is nearly zero. We now say that the **limit** of the average velocity Δx / Δt as Δt approaches zero is the instantaneous velocity. It is normally written as dx/dt and is known as the **derivative** of x.

Δt
I'm getting infinitesimally small...
Δx
...as am I.
How close is our increment to zero?
When does it actually become zero, and what happens then?
This is one of the deep mysteries of calculus.
In this specific connection, we conceive of zero not as a number but as an infinitely small magnitude.
We will return to it in due course!

Meanwhile, if we plot x as a graph against t, the derivative gives us the slope of the tangent to the curve at t.

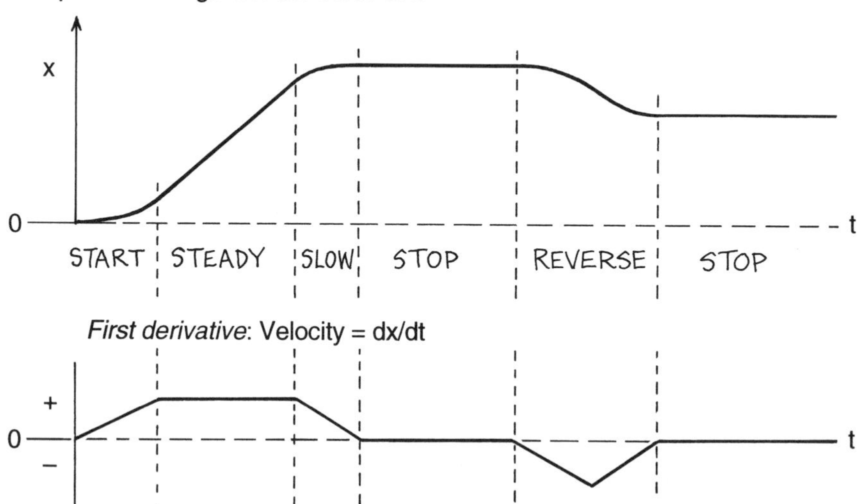
x
0
t
START
STEADY
SLOW
STOP
REVERSE
STOP
First derivative: Velocity = dx/dt
+
0
–
t

We can also take the derivative of the derivative to obtain the second derivative – in our example of the vehicle on the road, the second derivative will give us the rate of change of velocity, or acceleration.

Second derivative: Acceleration = d^2x/dt^2
+
0
–
t
Phew!... That was a bit complex, wasn't it?
Watch out, there's another bit of analysis coming up

Integration

Now we come to the masterstroke, the relationship that made "the calculus" the most powerful mathematical formalism ever.
Crucial to it were the two sorts of problems of the properties of curves, one involving the curve as a whole and the other involving the curve at a single point.
3
2
1
2
CHORD 1
CHORD 2
CHORD 3
TANGENT
Let's start with derivatives and their inverses.

The former had been solved by special methods of "exhaustion",

and the latter by drawing chords to the curve through the point.

Once curves were perceived as graphs of functions, then the problems of areas could be seen in a double perspective. On the one hand, areas could be "exhausted" by thin vertical strips; and on the other, the area **as a new function** is just the one whose derivative equals the original function. Then one single method, taking derivatives and taking their inverses, would solve both classes of problems.

We can see how it works in terms of our vehicle travelling along the road, and the three graphs for distance, velocity and acceleration. Instead of looking at the graphs starting with the distance function x(t) and going through the derivatives, let's start with the derivatives and work our way back to the distance function.

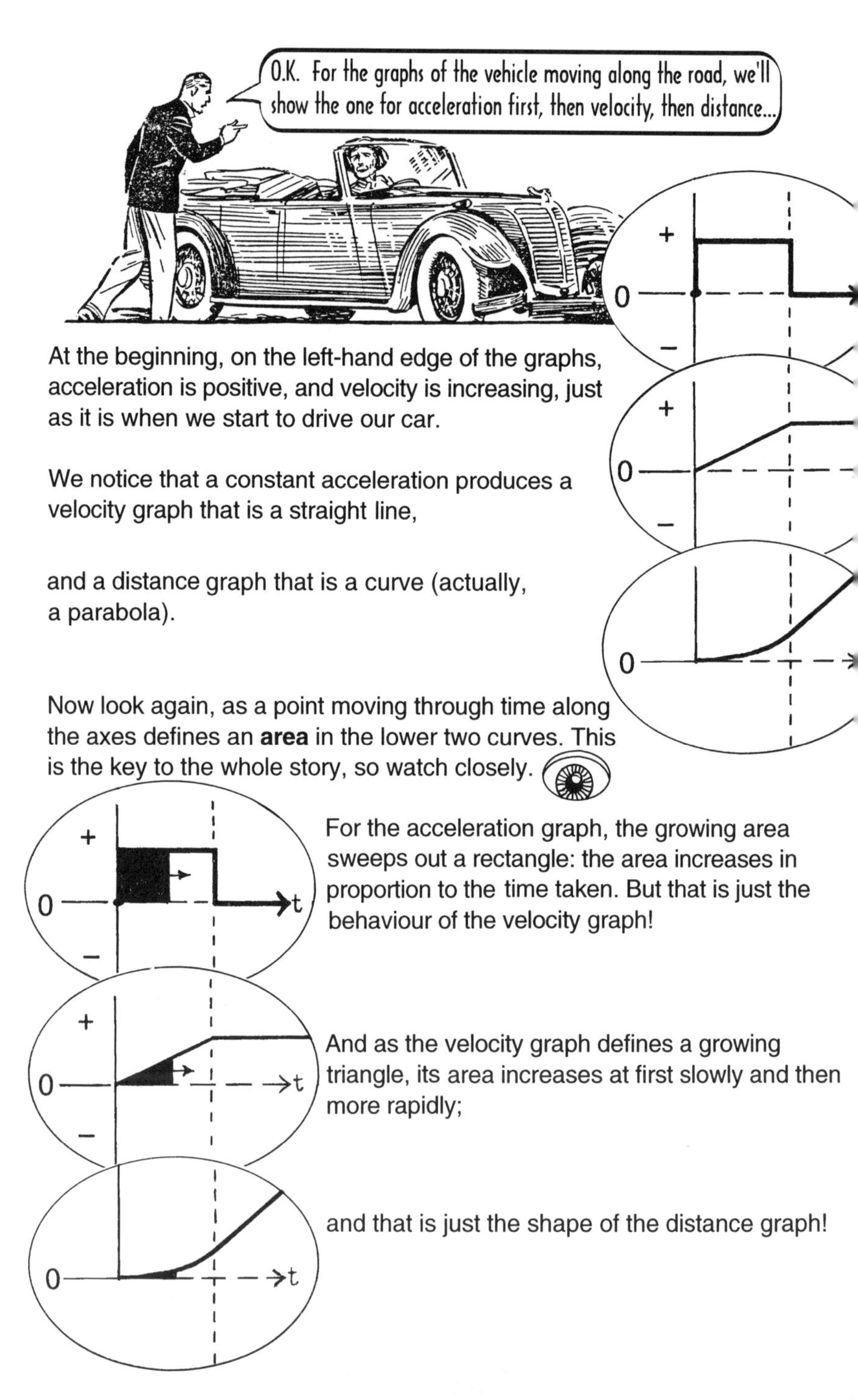

O.K. For the graphs of the vehicle moving along the road, we'll show the one for acceleration first, then velocity, then distance...
+
0
–
t

At the beginning, on the left-hand edge of the graphs, acceleration is positive, and velocity is increasing, just as it is when we start to drive our car.

We notice that a constant acceleration produces a velocity graph that is a straight line,

and a distance graph that is a curve (actually, a parabola).

Now look again, as a point moving through time along the axes defines an **area** in the lower two curves. This is the key to the whole story, so watch closely.

For the acceleration graph, the growing area sweeps out a rectangle: the area increases in proportion to the time taken. But that is just the behaviour of the velocity graph!

And as the velocity graph defines a growing triangle, its area increases at first slowly and then more rapidly;

and that is just the shape of the distance graph!

What we see from this is that if one function is the **derivative** (or rate-of-change) of another, then that other is the **area-function** of the first!

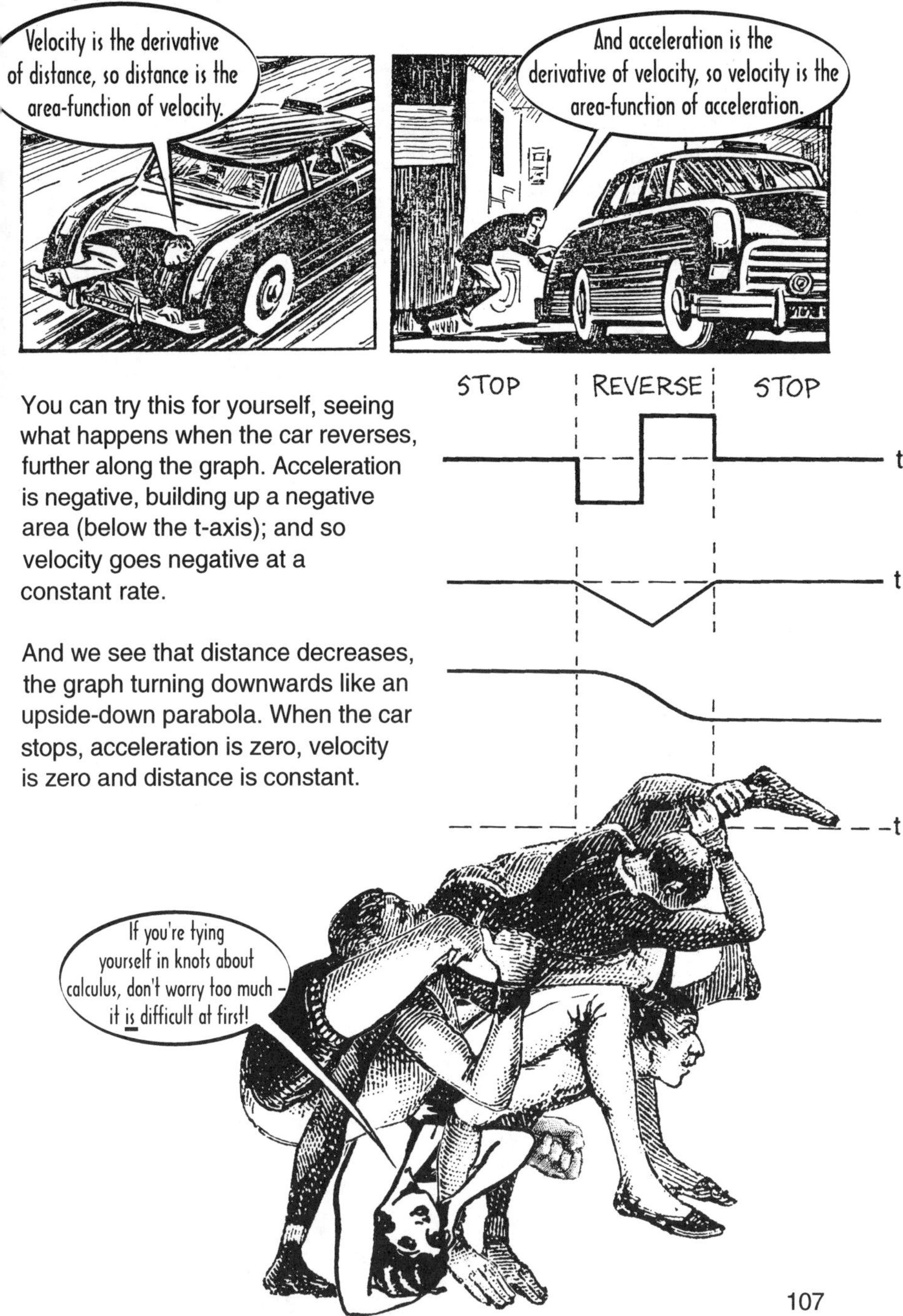

Velocity is the derivative of distance, so distance is the area-function of velocity.
And acceleration is the derivative of velocity, so velocity is the area-function of acceleration.
STOP
REVERSE
STOP
t
t
t
If you're tying yourself in knots about calculus, don't worry too much – it is difficult at first!

You can try this for yourself, seeing what happens when the car reverses, further along the graph. Acceleration is negative, building up a negative area (below the t-axis); and so velocity goes negative at a constant rate.

And we see that distance decreases, the graph turning downwards like an upside-down parabola. When the car stops, acceleration is zero, velocity is zero and distance is constant.

All we need to do now is to see how the other conception of integrals, that of area, fits in with that of inverse-derivative. That idea was how Newton conceived the calculus; while Leibniz started with areas as sums of infinitely thin strips.

Starting with a curve for velocity v(t), we imagine its area as filled up by very thin strips. Each has base Δt, and height v(t).

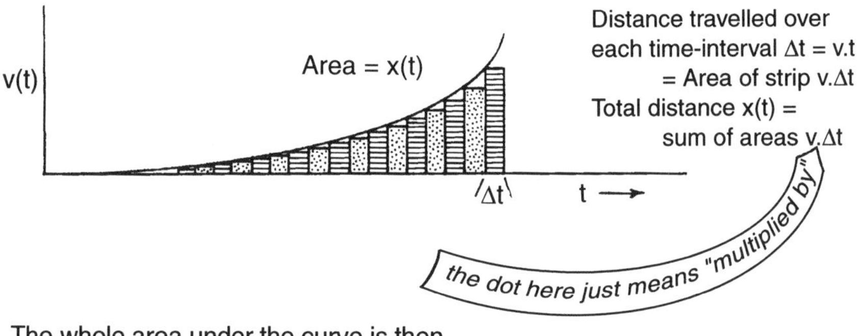
v(t)
Area = x(t)
Δt
t →
Distance travelled over each time-interval Δt = v.t
= Area of strip v.Δt
Total distance x(t) =
sum of areas v.Δt
the dot here just means "multiplied by"

The whole area under the curve is then

Sum {all the strips v(t).Δt}

Each of these areas describes a distance x, travelled at constant velocity v through time-interval t.

Now, say I, let the intervals become infinitesimally small and we can smooth them to have base dt; the sum has a special symbol...
LEIBNIZ
∫v(t)dt

To get back to the inverse-derivative relation, all we need is to imagine the "last" thin strip, which is just Δx itself.

Then since

$$\Delta x = v.\Delta t$$

we have $$\Delta x/\Delta t = [v.\Delta t]/\Delta t$$

and so $$dx/dt = v(t)$$

So the derivative of the integral defined as the sum of strips, is just that same function whose area produced the integral.

It is (relatively) easy to find the derivatives of functions given algebraically or in terms of some special functions. To find the algebraic form of the area-function, we just find that particular function whose derivative is the original function. Problems of properties of curves as a whole are reduced to the simpler problems of properties of curves at a point.

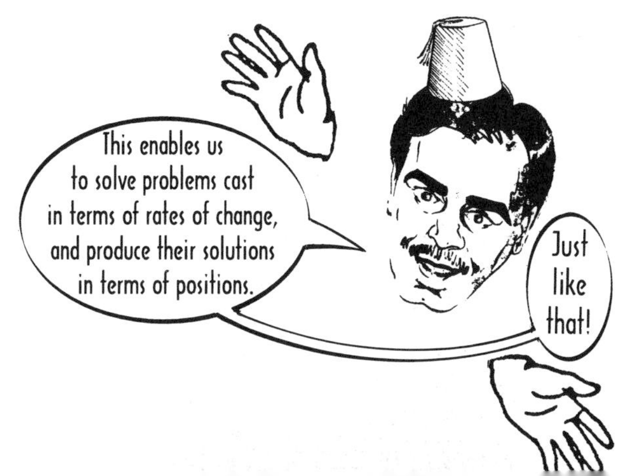
This enables us to solve problems cast in terms of rates of change, and produce their solutions in terms of positions.
Just like that!

If some of this seems strange and hard to follow, don't worry.
Calculus is full of paradoxes which take some getting used to.

Calculus was first applied to mechanics and astronomy. The techniques of differential equations enabled the creation of mathematical physics. Only then could we have sciences of heat, energy, electricity and magnetism. The world of modern science, underpinning the world of modern technology, depends quite directly on the calculus.

Berkeley's Questions

What about that increment, and the mystery of how it got to zero? People asked that of Leibniz and Newton at the time, and got unsatisfactory answers. Then, the Irish philosopher and Anglican Bishop, **George Berkeley** (1685-1753), raised the questions in a very sharp form.

I observe that forming a quotient with the increments makes sense only if it is not zero; otherwise we are dividing by zero, and that is illegitimate.
Is the increment always non-zero, or does it become exactly zero, or is it "the ghost of a vanished quantity"?
And apart from that, sirrah, Mr Newton is naked.

From William Blake's *Newton*

Berkeley's purpose was to show that "freethinkers", who claimed that Science and Reason would soon replace the mysteries and superstitions of religious belief, were as obscure and dogmatic as the worst of the theologians. In the subtitle of his pamphlet, he asked: "... whether the object, principles, and inferences of the modern Analysis are more distinctly conceived, or more evidently deduced, than religious Mysteries and points of Faith". For him, the answer was clearly...

NO!

Some mathematicians attempted to answer Berkeley's pamphlet, *The Analyst*. He used their answers to expose their confusions ever more cruelly. His reply, *A Defence of Freethinking in Mathematics*, is a masterpiece of critical analysis.

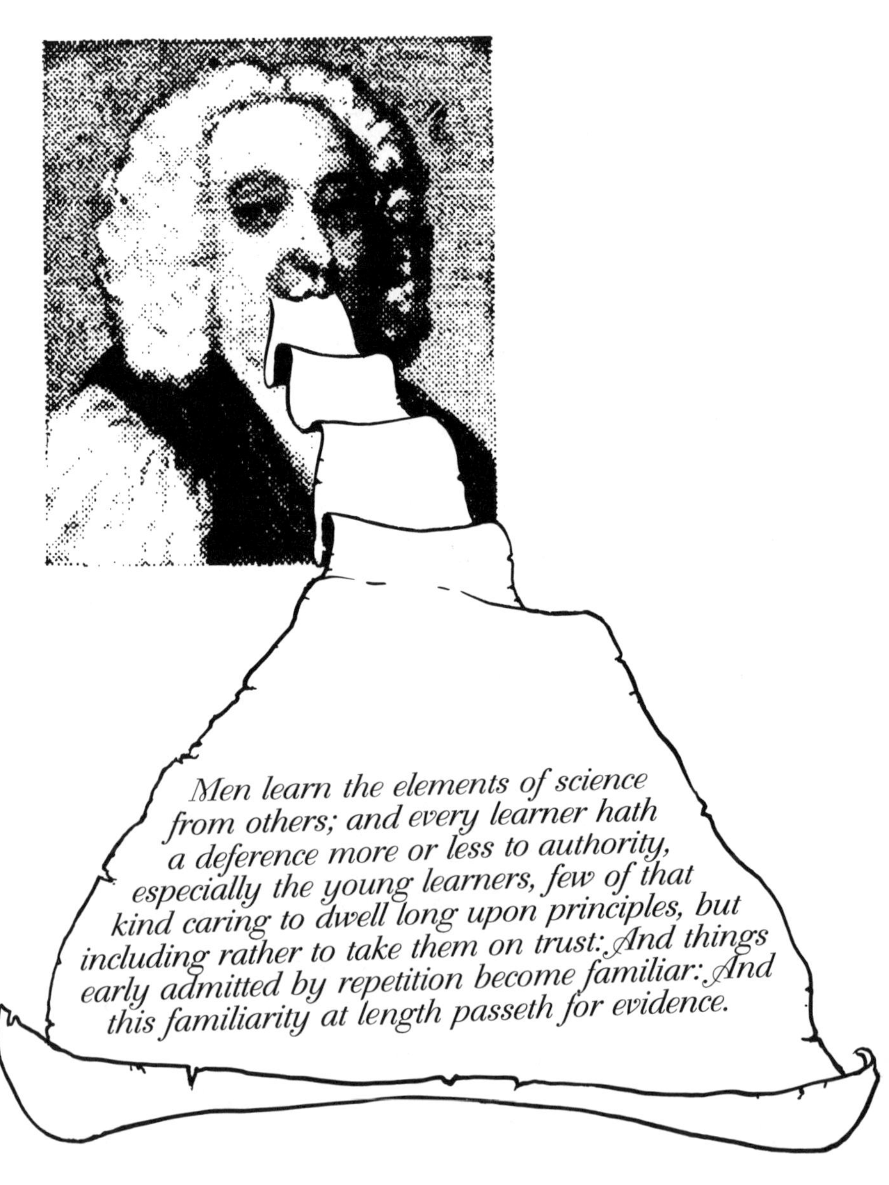
Men learn the elements of science from others; and every learner hath a deference more or less to authority, especially the young learners, few of that kind caring to dwell long upon principles, but including rather to take them on trust: And things early admitted by repetition become familiar: And this familiarity at length passeth for evidence.

Berkeley was arguing that learning how to solve problems in mathematics and science does not necessarily help us to understand what they are all about. He anticipated the image of scientific research developed by **T.S. Kuhn** (1922-95) who described "normal science" as a practice of "puzzle-solving" within a "paradigm" (framework of thought) that is unquestioned, and indeed unquestionable, so long as it works. For Kuhn, ordinary science is actually quite a narrow-minded practice, and science teaching (including mathematics) is necessarily rather dogmatic.

Please sir, are we dividing by zero?
Mathematics teachers sometimes make a joke about "proof by intimidation".
Ha blooming ha...
This is nowhere more necessary than in rationalizing the mysteries of the calculus.

Euler's God

It was the Swiss mathematician **Leonhard Euler** (1707-83) who first linked the exponential and trigonometric functions together and produced a formula of their relationship.

Euler had an extraordinary mathematical genius, and stories of his prowess are legion. He was employed at the court of Frederick the Great of Prussia, where he encountered the French encyclopedist and philosopher **Denis Diderot** (1713-84). Diderot was a hard-headed atheist...

I challenge the pious Euler to produce a mathematical proof of the existence of God.
Sir, $(a + b^n) / n = x$, hence God exists. Reply!

Diderot was dumbfounded, and fled back to the safety of the Paris *salons*.

The formula mentioned in that story has nothing special about it. But Euler also developed one of the most beautiful formulae in all mathematics, that certainly does make one stop and think.

Euler's formula is a mysterious, transcendent expression that connects the five most fundamental numbers in the universe:

$e^{\pi\sqrt{-1}} + 1 = 0$

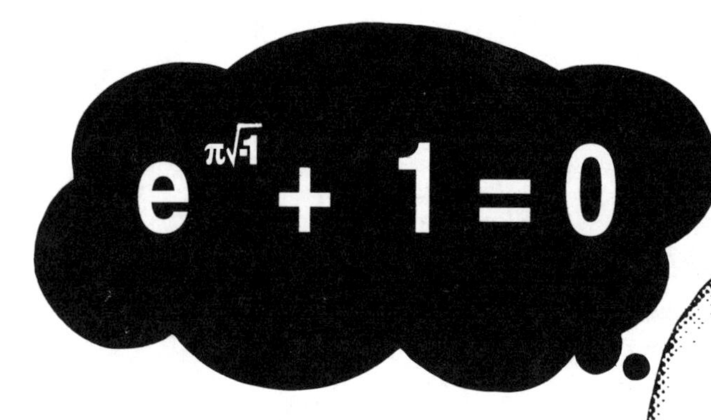
$e^{\pi\sqrt{-1}} + 1 = 0$

Looking at them in reverse order, the first to appear is 0, the mysterious quasi-number. Then we find 1, the unit, the foundation of all numbers. It also appears as its negative, encased in the square-root ($\sqrt{-1}$, called "i"). It is the basic unit in the "imaginary numbers" that have fascinated so many cultures and civilizations. Then there is the oldest mathematical constant, π, measuring the ratio of the circumference of a circle to its diameter. The last number, the most recently discovered, is the transcendental number e, the base "natural" exponential growth.

Could a relationship like this possibly have been discovered by experiments, however long repeated?

In fact, Euler's divine formula comes from a function he discovered, that links complex numbers with the trigonometric functions discovered by the Muslim mathematicians (see page 85).

We saw that the function e^x has a very rapidly growing graph (see page 99). By contrast, the graph of $e^{\sqrt{(-1)}x}$ is a circle! Its radius is just one unit; and x is the angle made by the line from the origin to the point. As the point goes around the circle, x increases from 0 to 2π. But if we look at that graph with an eye to the trigonometric functions, we see that the "Real" part of the number $e^{\sqrt{(-1)}x}$ is just cos x, and the "Imaginary" part is sin x.

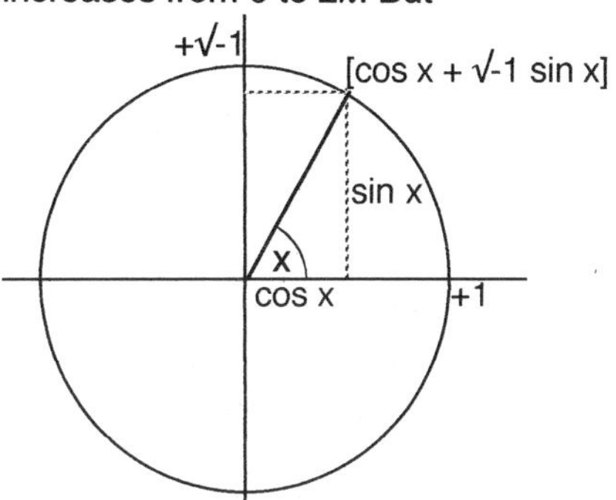
+√-1
[cos x + √-1 sin x]
sin x
x
cos x
+1

So we can write:

$e^{ix} = \cos x + i \sin x$,
where i is the common symbol for $\sqrt{-1}$.

What if the point goes around the circle a second time, so that x keeps on increasing? The functions e^{ix}, cos x and sin x just keep on repeating themselves. They are said to be **periodic**. The graph of y = sin x looks like this:

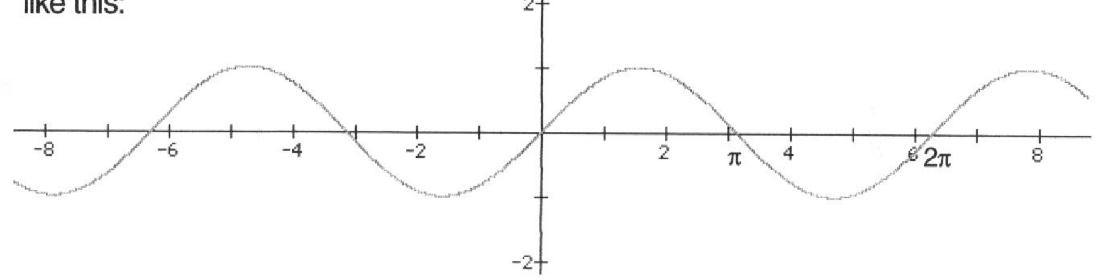
2
-8
-6
-4
-2
2
π
4
6
2π
8
-2

That is like many phenomena that are either alternating in time, like electric current, or are waves propagating in space, like sound. The sines and cosines are the building blocks of all the complex wave forms that carry messages. And in doing mathematics with them, using the "imaginary exponential" form converts cumbersome special calculations into neat and easy exercises.

So, my divine formula does a lot of work in the world of technology and industry!

Non-Euclidean Geometries

We have seen that Euclid deduced all his geometry from a few "common notions" and self-evident "postulates". But one of these, about parallel lines, seems more like a theorem. It was an embarrassment for centuries, for it cast doubt on the truth and perfection of Euclid's system. Then suddenly it became the basis for a great leap in the mathematical imagination: the invention of non-Euclidean geometry.

This was accomplished by several people. But the first one didn't know he was doing it! This was a Jesuit mathematician, G. Saccheri, who was determined to finish all the quibbling, once and for all. In his book *Euclid Cleared of Every Blemish* of 1733, he tried to show that it would be impossible to do geometry without the parallel postulate.

I actually proved some theorems... but they were ridiculous, and I stopped.
This was the greatest own-goal in the history of scientific thought.

There was nothing wrong with the results, and they were repeated later by the real inventors, who knew what they were doing.

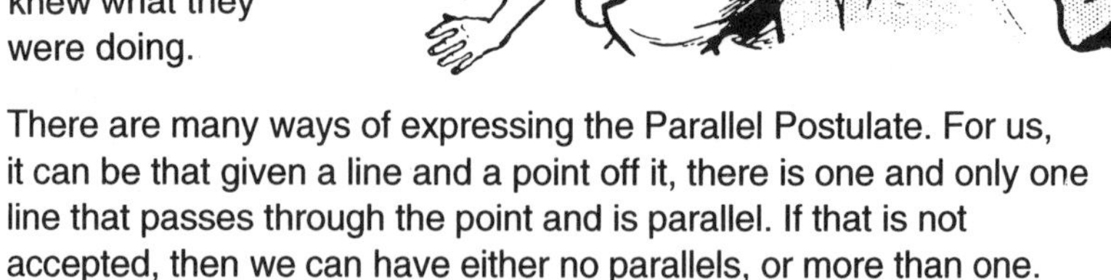

There are many ways of expressing the Parallel Postulate. For us, it can be that given a line and a point off it, there is one and only one line that passes through the point and is parallel. If that is not accepted, then we can have either no parallels, or more than one.

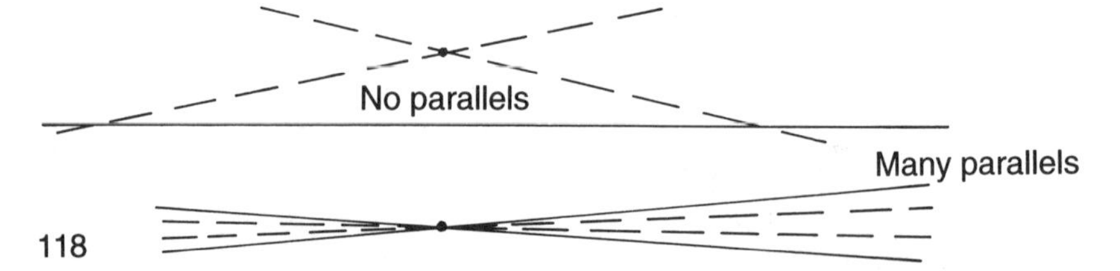
No parallels
Many parallels

First, the case of **many** parallels was discovered, nearly simultaneously, by two mathematicians, the Hungarian **Janos Bolyai** (1806-60) and the Russian **Nikolai Lobachevski** (1792-1856). Later, the German **Georg Riemann** (1826-66) worked on the case of **no** parallels. Then it was realised that these geometries can be done by constructions on special sorts of surfaces.

For Riemann's geometry, a sphere is a good example, if we understand "line" as a great-circle. That is the curve on the surface made by a plane that passes through the centre of the sphere. (See page 84 for spherical trigonometry.) Since any two great-circles must intersect twice, there are no parallels.

Lobachevski
Bolyai
For our geometry, the surface is not quite so easy to visualize.
It is like a trumpet-shape, formed by a curve rotated about a line.

Here, we think of a "line" as the shortest path between two points. And it turns out that there are many "parallels", lines that never meet the given one.

As people got used to the idea of non-Euclidean geometries, it weakened the faith that mathematics tells us truths that are logical and infallible. But it took a long time for that revolutionary thought to sink in.

N-Dimension Spaces

Another counter-intuitive development within geometry was the study of spaces of more than three dimensions. An extension of Descartes' system of algebraic geometry to more dimensions is quite straightforward. Instead of being located on a plane by the coordinates (x, y), a point in "hyperspace" might have coordinates (x_1, x_2, x_3 … x_n). Of course, the properties of curves in such hyperspaces are very different from those in two or three dimensions. But conceiving many dimensions now seems to present no difficulty to us.

We do it all the time on 'Star Trek'
To boldly go into hyperspace...
I come from another dimension.
2D or not 2D?
I'm working flat out on these drawings...

In Victorian times it was very different.

A little masterpiece of mathematical fiction and social criticism was written about this, called *Flatland*. It describes a society where the people are polygons living on a plane. Like the Victorians, they are obsessed with status, which depends on the number of a person's sides. Gentry have four, aristocrats many, workers three, and women are just a needle!

The hero, "A Square", has an experience of three dimensions, through his friendship with a Sphere. This higher being appears to the flatlanders every five hundred years, manifesting as a circle that starts as a point, then grows larger, then diminishes and finally vanishes. What is incomprehensible to the flatlanders is simply the Sphere passing through their plane. The Sphere befriends our Square, and takes him on a journey through space. He shows him "lineland" and "pointland", which contain some very self-satisfied creatures. He also enables him to peer into the private lives of the flatlanders. But on his return to the plane, the Square suffers badly. He tries to describe Space, but how can he show "up" to his friends? They think him deranged.

"O day and night, but this is wondrous strange"

FLATLAND

A ROMANCE OF MANY DIMENSIONS

No Dimensions — POINTLAND

One Dimension — LINELAND

Two Dimensions — FLATLAND

Three Dimensions — SPACELAND

With Illustrations by the Author, A SQUARE

(EDWIN A. ABBOTT (*1838-1926*))

Finally I become disillusioned about the enlightenment of higher-dimensional beings!

Evariste Galois

Throughout the 19th century, algebra experienced a vast increase in power, generality and abstraction. It became more and more rooted in **formalism**. Gradually, the idea emerged that the system of formalisms could refer to other things than numbers and their arithmetical operations.

A great step forward in this was made by the French mathematician **Evariste Galois** (1811-32), undoubtedly one of the most tragic figures in the history of mathematics. He was an ardent Republican in a period of severe political reaction. He may have been the victim of official *agents-provocateurs*, who arranged a love affair between the hapless youth and the *fiancée* of a renowned duellist. Galois was killed at the tender age of twenty-one. On the last night of his life, he wrote out a manuscript containing all his ideas. This was nearly lost, but was eventually recovered and published some fifteen years later.

Galois attacked an old problem, that of finding a solution in square-roots for the general quintic equation $x^5 + \ldots = 0$. By his time, there was a consensus that the task was impossible, but no one had proved it.

That is what I set out to do, and in the course of developing my arguments, I hit on a new idea – the concept of Groups.
Very far-sighted of him.

Groups

I have something I want to share with group ...
My feet are going to sleep.

Groups are mathematical structures defined by elements and rules of combination. They can be thought of as systems of arithmetic without numbers. Their elements need have no relationship to counting or measuring, and are not "numbers" in the normal sense of the word. Galois realised that there can be sequences of operations that behave as if they are additions.

These sequences have just a few defining properties.

1. For any two elements, there is a third resulting from their combination, such as: 2 + 2 = 4

2. There is an "identity" element, which does not change the element it combines with, as in 2 + 0 = 2

3. And for any element there is an "inverse", which, when combined, yields the identity, thus: 2 + (-2) = 0

This is what is called a group
Those who have been exposed to the "new maths" at school know it well!

For an example of a group, which is actually a very simple version of what Galois did, we consider a set of four objects named a b c d

These are not the elements of the group. The group consists of the operations of cycling of these four objects. We imagine a cycling among them, either by one place, giving b c d a

or by two, giving c d a b

or by three, giving

d a b c

If we cycle by four places, we are back where we started, and so that is the identity.
A lousy pun on 'cycle'.

We might call these cyclings A, B, C and I. Then A + C amounts to a cycling of 1 + 3 places, which gives 4 places, or Identity! It's easy to do a full "additions table" on this set of just four elements: three plus an identity.

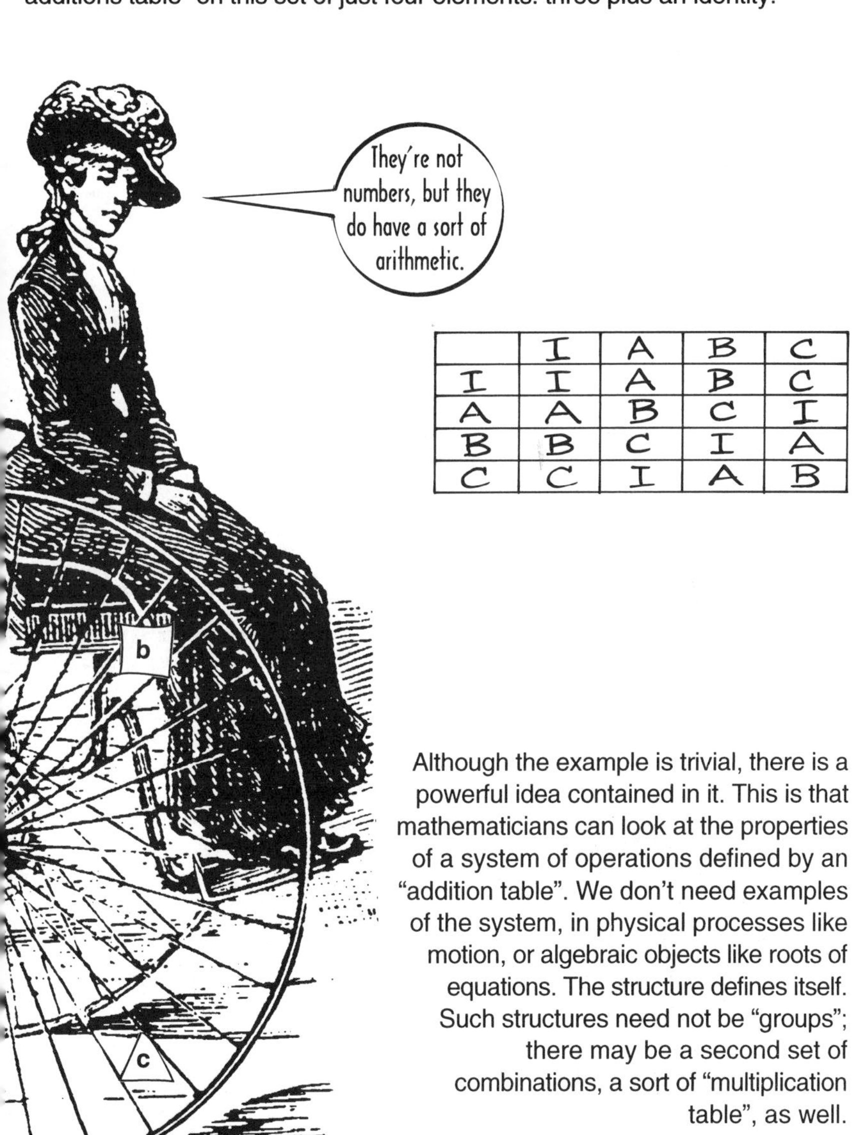
They're not numbers, but they do have a sort of arithmetic.
b
c

	I	A	B	C
I	I	A	B	C
A	A	B	C	I
B	B	C	I	A
C	C	I	A	B

Although the example is trivial, there is a powerful idea contained in it. This is that mathematicians can look at the properties of a system of operations defined by an "addition table". We don't need examples of the system, in physical processes like motion, or algebraic objects like roots of equations. The structure defines itself. Such structures need not be "groups"; there may be a second set of combinations, a sort of "multiplication table", as well.

Boolean Algebra

Soon, other sorts of operations were looked at. One of the most exciting was developed by the British mathematician **George Boole** (1815-64), which allowed mathematical methods to be applied to such non-quantifiable entities as logical propositions.

In a modern form, this is the algebra of the combination of sets, or "Boolean algebra".

I modestly called my efforts "the Laws of Thought".
This has the operations of "union" (the resulting set has all the members belonging to either)...
... I'd prefer not to lose one of my members during this operation, if that's O.K. ...
...and "intersection" (the combined set has only the members belonging to both).
Boolean algebra comes into play whenever we have to make a choice among options. It is there when we do a search on the Internet.

Suppose we are looking for a recipe for "Hot Cross Buns". So we type in the keywords:

HOT CROSS BUNS

The search engine then asks us, do we want sites with

Any of the keywords or **All** of the keywords.

The first choice will give us all the sites which have either Hot or Cross or Buns. In terms of Venn diagrams, that is:

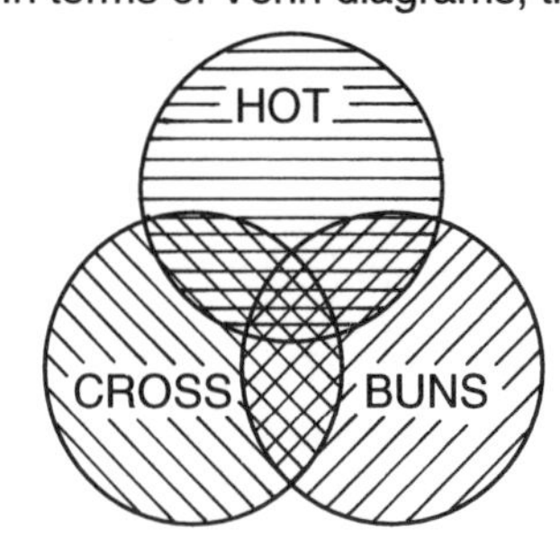
HOT
CROSS
BUNS

In terms of the sets, that is {Hot} + {Cross} + {Buns}. That will be lots of sites, with lots of interesting but irrelevant entries.

But if we want only "Hot Cross Buns" and nothing else, then we get the sites which have only Hot and Cross and Buns. The picture then is:

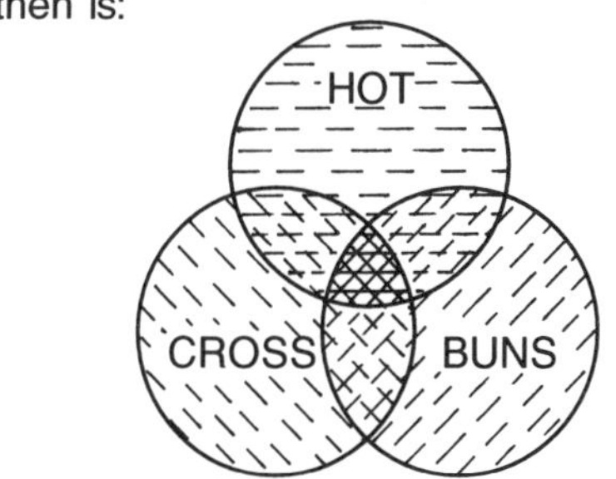
HOT
CROSS
BUNS

In terms of the sets, that is {Hot} X {Cross} X {Buns}. So then we get Hot Cross Buns and nothing else.

Since computer programs include many operations of choosing, not just doing arithmetic with numbers
–(in French they are called "ordinateurs")–
Boolean algebra is fundamental to their design.

The "arithmetic" of Boolean algebra is interesting because, unlike in ordinary arithmetic, we have both of the "distributive" relationships:

A X (B + C) = (A X C)   and also   A + (B X C) = (A + C)

In ordinary arithmetic the first works, the second doesn't. But for sets where 'X' is intersection and '+' is union, both work as this illustration called a "Venn diagram" shows.

Here's the "distributive law" that works as with numbers:

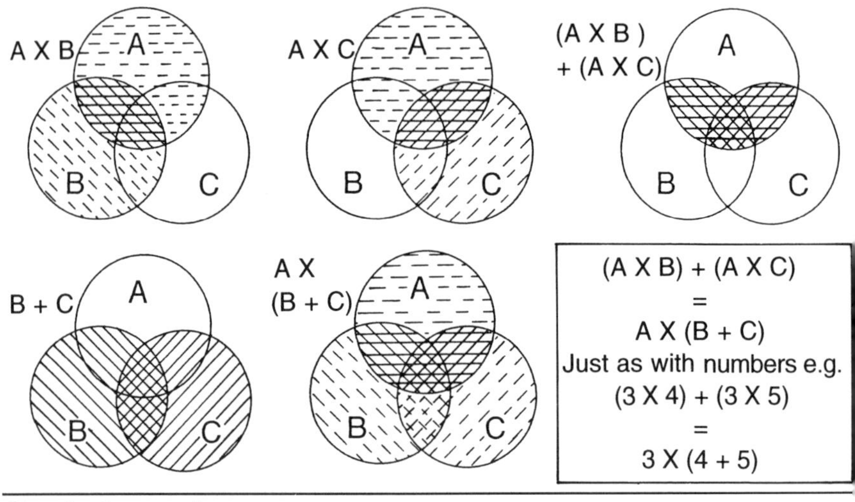
A X B
A
B
C
A X C
A
B
C
(A X B)
+ (A X C)
A
B
C
B + C
A
B
C
A X
(B + C)
A
B
C
(A X B) + (A X C)
=
A X (B + C)
Just as with numbers e.g.
(3 X 4) + (3 X 5)
=
3 X (4 + 5)

Now for the surprise...

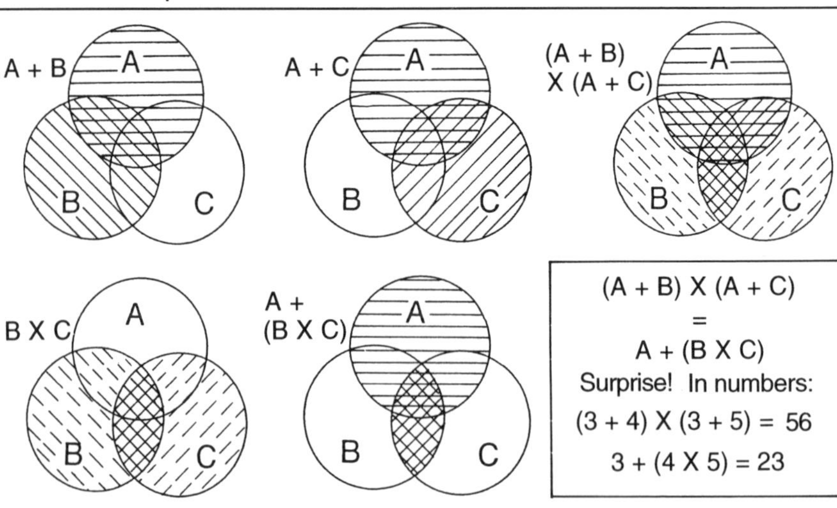
A + B
A
B
C
A + C
A
B
C
(A + B)
X (A + C)
A
B
C
B X C
A
B
C
A +
(B X C)
A
B
C
(A + B) X (A + C)
=
A + (B X C)
Surprise! In numbers:
(3 + 4) X (3 + 5) = 56
3 + (4 X 5) = 23

Examples such as these gave mathematicians great scope for their imaginations. The "arithmetic" studied by mathematicians increasingly became quite different from the one we know from numbers.

Cantor and Sets

While some were worrying about numbers, others were concerned with the infinite. Sets that are actually-infinite had previously been left to speculation, mathematical and mystical. The German mathematician **George Cantor** (1845-1918) went boldly forth to tame the infinite.

I showed how to construct various such sets, and I also proceeded to count them.

He provided a scheme for counting all the fractional numbers, laying them out in a pattern like this.

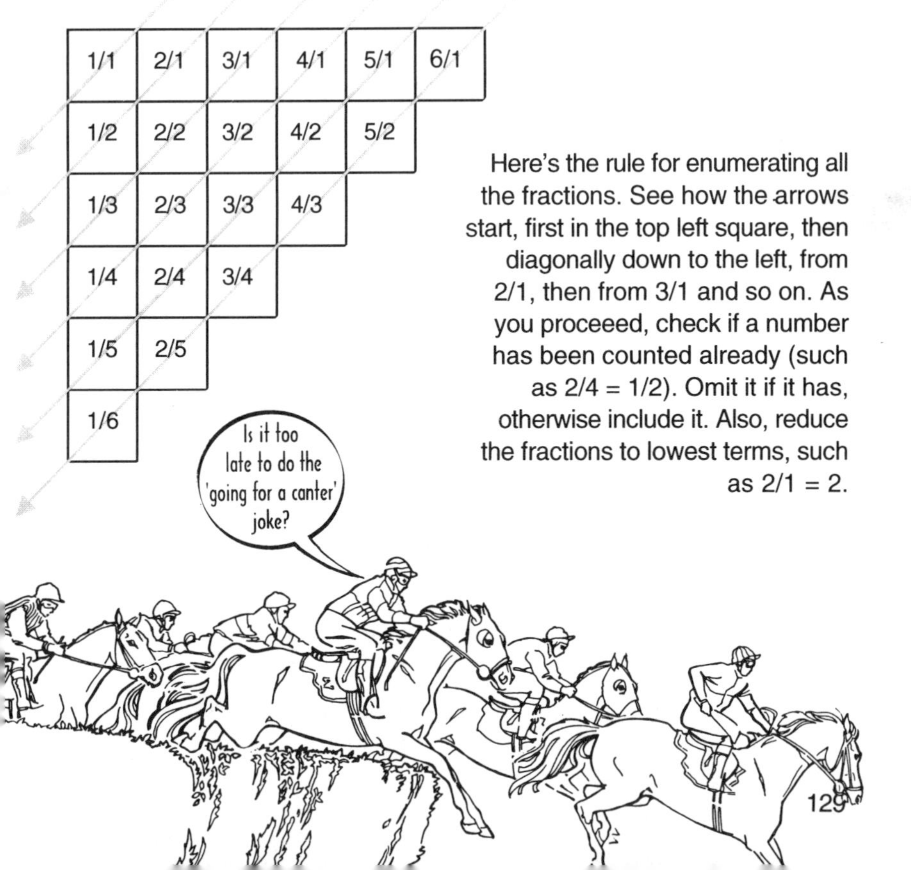
1/1 2/1 3/1 4/1 5/1 6/1
1/2 2/2 3/2 4/2 5/2
1/3 2/3 3/3 4/3
1/4 2/4 3/4
1/5 2/5
1/6
Is it too late to do the 'going for a canter' joke?

Here's the rule for enumerating all the fractions. See how the arrows start, first in the top left square, then diagonally down to the left, from 2/1, then from 3/1 and so on. As you proceeed, check if a number has been counted already (such as 2/4 = 1/2). Omit it if it has, otherwise include it. Also, reduce the fractions to lowest terms, such as 2/1 = 2.

Sets...
tennis...
geddit?
Will you button it? I'm trying to follow the maths!

We then have the sequence:

1, 2, 1/2, 3, 1/3, 4, 3/2, 2/3, 1/4, 5 ...

You can see that this is the same as looking at all the fractions (including integers) whose numerator and denominator total to 2, then to 3, then to 4 and so on, and starting with the largest numerator each time. Any number, integer or fraction will be reached sooner or later.

Similarly, all the numbers that are solutions of algebraic equations, like √2 and √(-1), can be enumerated.

I wonder if there is an enumeration for all of them?
But there are the extras, like π and e, and doubtless many, many more.

Cantor's work actually proved the opposite of his intention. He found that the set of all "real numbers", the points on a line, cannot be enumerated. His proof takes only a few lines, but you have to watch closely!

Suppose that we have all the numbers enumerated, just like the fractions and algebraic numbers. Then there will be a list of them, infinitely long like the list we constructed just previously. Now, just as with the fractions, the numbers on that list don't appear in order of size.

For simplicity, we take all the numbers between zero and one, and lay out their decimal expansions. The list might look something like:

$N_1 = .7166932.....$
$N_2 = .4225896.....$
$N_3 = .7796419.....$
$N_4 = .3228952.....$
......

These numbers were arbitrarily chosen!

The dots at the end of each of the strings of digits indicate that the string goes on indefinitely.

And the line of dots after N_4 indicates that the sequence of numbers N also goes on indefinitely.

Now, if all the "real numbers" are included in that list, then any number that we construct out of those we see there, will also be somewhere in the list.

If not, then we would have to admit that we have not included them all.

CANTOR OIL

How could we possibly construct a number that is not on that list? Well, suppose we have one that is different from the first number in the first place, different from the second number in the second place, third in the third, fourth in the fourth, and so on, and on. We can construct such a number just by having the digit in each place to be one more than the digit of the number in the list.

For the list that we have made we find...

1st place:	7 → 8
2nd place:	2 → 3
3rd place:	9 → 0
4th place:	8 → 9
....	

As you can see, the actual numbers we put down don't matter. They could be completely different; the argument is unaffected.

So that new number, that we might call "strange", is now (for the list we have given)
S = .8309.....

Here is the punch-line...

Where is S on the list?

Not in the first place, not the second, nor the third, nor ... anywhere!

So our assumption that we could enumerate all the real numbers is FALSE.

Cantor was working with two levels of infinity: the denumerables (like the ordinary numbers) and the points on a line. How were they to be related? Then he got a method for generating and describing higher orders of infinity! For this, we use the idea of "subset". If we have a set of three elements, a, b, and c, then its subsets are the pairs ab, ac, bc, and the single elements a, b, and c; and (by convention) the "empty" set (the one with no members) and also the original set itself.

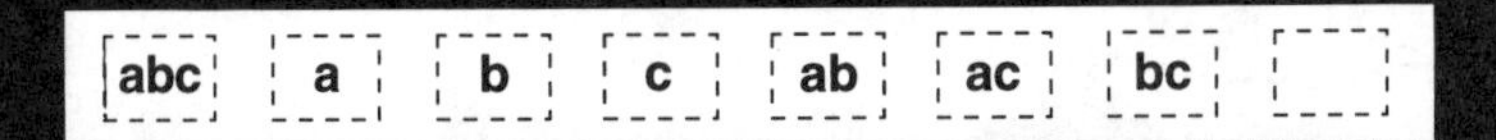

abc
a
b
c
ab
ac
bc

Counting them up, we can see that this totals eight elements, or 2^3. This new set is called the **power set** of the original one; and if the original one has N elements, the power set has 2^N.

Now Cantor could generate ever-bigger sets, just making power sets, one after another. He produced a new symbol for the "size" of these sets. Or rather, being Jewish, he adopted the old Hebrew letter "Aleph", or $\aleph$. So, if the denumerable sets are of size Aleph-null, or $\aleph_0$, their power set is $2^{\aleph_0}$ and so on.

On the other hand, the set of the real numbers on the line, the first denumerable set, is $\aleph_1$.

It would seem reasonable to suppose that $2^{\aleph_0}$ is just equal to $\aleph_1$, but that hypothesis tantalized mathematicians for generations afterwards.

IMPOSSIBLE
Roaming around infinities like these was exciting, indeed bewildering; but then disaster struck!!!

For when one is talking about "sets" in such a general way, there is nothing to stop one from referring to "the set of all sets" – it makes grammatical sense, doesn't it? Now, that must be the biggest set of all, and its size will be a certain Aleph, let's call it $\aleph_F$, for final. But, like any other set, it will have a power set, whose number can be defined as

$$2^{\aleph_F}$$

and this is certainly greater than $\aleph_F$. So, what we defined as the truly biggest set, the set of all sets, can generate an even bigger one. The idea is self-contradictory!

It is like the revenge of all those children whose teachers put them down when they asked about the last number.

Crisis in Mathematics

The paradoxes of the infinite discovered by Cantor presented a new sort of challenge to mathematics. These were no longer cases where the mathematical objects seemed to defy intuition, like $\sqrt{-1}$ or dx/dt. Rather, they were plainly self-contradictory. Yet they had been derived by arguments no different in detail from those of conventional mathematics.

Mathematics was in a state of crisis.
Is mathematics destroying its own foundations?

At the beginning of the 20th century, a host of philosophers and mathematicians set out to resolve the crisis. They asked...

Russell and Mathematical Truth

Amongst those who were keen to resolve the crisis was **Bertrand Russell** (1872-1970). His long career included logic, philosophy, progressive education, and finally civil disobedience in protest against nuclear weapons. To him, mathematics represented the only genuine truth in the world, in opposition to the spurious claims of religion.

I (and others) studied <u>logical paradoxes</u> for clues to what went wrong with Cantor's analysis.

These had been known from classical Greek times. Some depended on the use of "all", as in "the set of all sets".

Others depended on self-reference, such as the statement...
..."This statement is false"

If the statement in quotes is true, then by its content it is false;...
...but if the statement in quotes is false, then by its content it is true!

One of the most ingenious of the paradoxes is about naming. Let us define "B" as "the least integer not nameable in fewer than nineteen syllables". In the ordinary way, it would be rather a large number to need nineteen syllables for its name: "seven hundred thousand million billion" only needs ten.

But the paradox is that this definition of "B" itself has only eighteen syllables!
(Count them!)
So "B" is nameable in fewer than nineteen syllables!
The name is an odd one, but never mind; it is a name, and it contradicts itself.

This paradox is very serious indeed, for it involves neither self-reference nor universality. It shows how difficult it would be to salvage certainty in mathematics by cleaning up its logical foundations.

So the campaign was eventually abandoned, even by Russell himself.

The only way out is to ban such self-referring statements.

But legislation for such "straight thinking" is not easy to frame...
...and other sorts of paradoxes kept cropping up.

Another line of attack was developed, in a last attempt to secure mathematical truth.
This was to consider mathematical arguments as pure formalisms, collections of symbols, and to see whether they could be shown to be rigorous in that way.
However this programme was soon exploded by one of its most brilliant recruits, me, Kurt Gödel.

A "proof" would be a set of lines of symbols, which were connected by transformation rules. The task was to show that "valid" proofs could be distinguished from invalid ones, so that any mathematical statement could be shown to be either true or false.

Gödel's Theorem

Kurt Gödel (1906-78) published his theorem in 1931 in response to **A.N. Whitehead** (1861-1947) and Russell's three-volume work on symbolic logic, *Principia Mathematica* (1910-13).

My theorem proved that any consistent mathematical system must be incomplete ...
...that is, in any system, formulae can be constructed that can be neither proved nor disproved within that system.
Furthermore, no mathematical system can be proved consistent without recourse to axioms beyond that system.
Given a finite number of axioms and rules for deducing axioms from them, we will always, if the system is consistent ...
...be able to produce at least one true statement that the system cannot prove.

His device was to use numbers in a new way. He assigned a number to every part of a mathematical statement, and then combined them to give a unique number to each statement. Then, by an argument reminiscent of Cantor's, he produced a "monster" number representing a statement, which was fully meaningful, but which could neither be proved nor disproved.

Gödel's theorem finished off, once and for all, the dream that mathematics could be an edifice of truths, all logically connected.

The Turing Machine

Power of a different sort came from the magnificent destruction wrought by Gödel. For the idea of generating mathematical statements in his completely abstract way was picked up by **Alan Turing** (1912-54).

In my hands, it became the specification of a computer which was a completely different thing from a mechanical calculator.

A "Turing machine" consisted of a tape and a program which responded to information on each successive section of the tape, performing the most elementary operations. Given the technology of the 1930s, this conception had no practical use whatever. But it furnished Turing with a version of Gödel's methods which he wanted for his research.

Quite soon, Turing's insights became very practical, as they guided the development of computers during the Second World War. These started as huge calculating machines, where the program is set up from the outside (by settings on knobs and switches). The big change came when the program was installed *inside* the computer as a special file, one that directs operations on all the others. Now there was no limit to its complexity or adaptability.

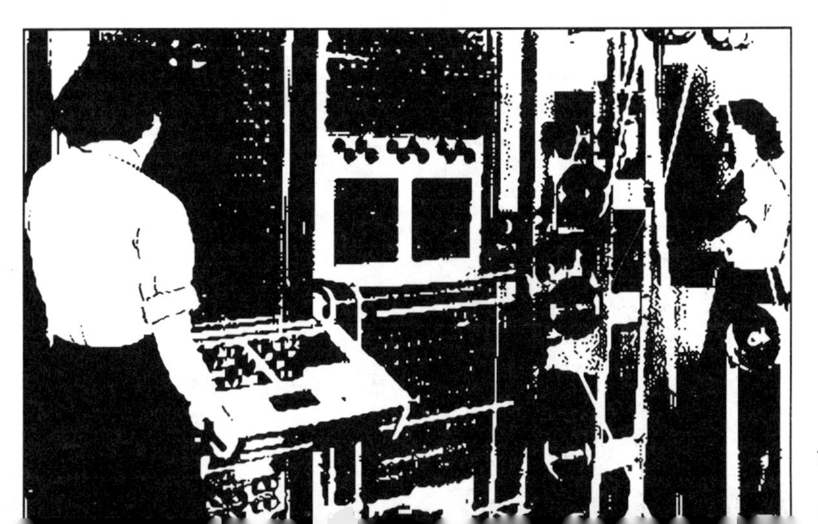

Turing himself helped to win the war by being on the team that broke the code of the German “Enigma” cipher machine. But he died tragically, almost certainly as a result of being persecuted (and prosecuted) as a homosexual. He was found with cyanide poisoning, and by his side was the poisoned apple with a bite taken out.

In its own way, Turing’s vision of an abstract computer turned out to be partly misleading in the long run. In his scheme of simple operations, there was no place for programming errors, or the need for “debugging”. For decades, computers were believed to be infallible; any mistakes were the result of human error. Only now, with the discovery of the “Millennium Bug”, do we all realize that the abstract, formal systems of computer theory and computer programs are not divine truths, but all-too-human productions.

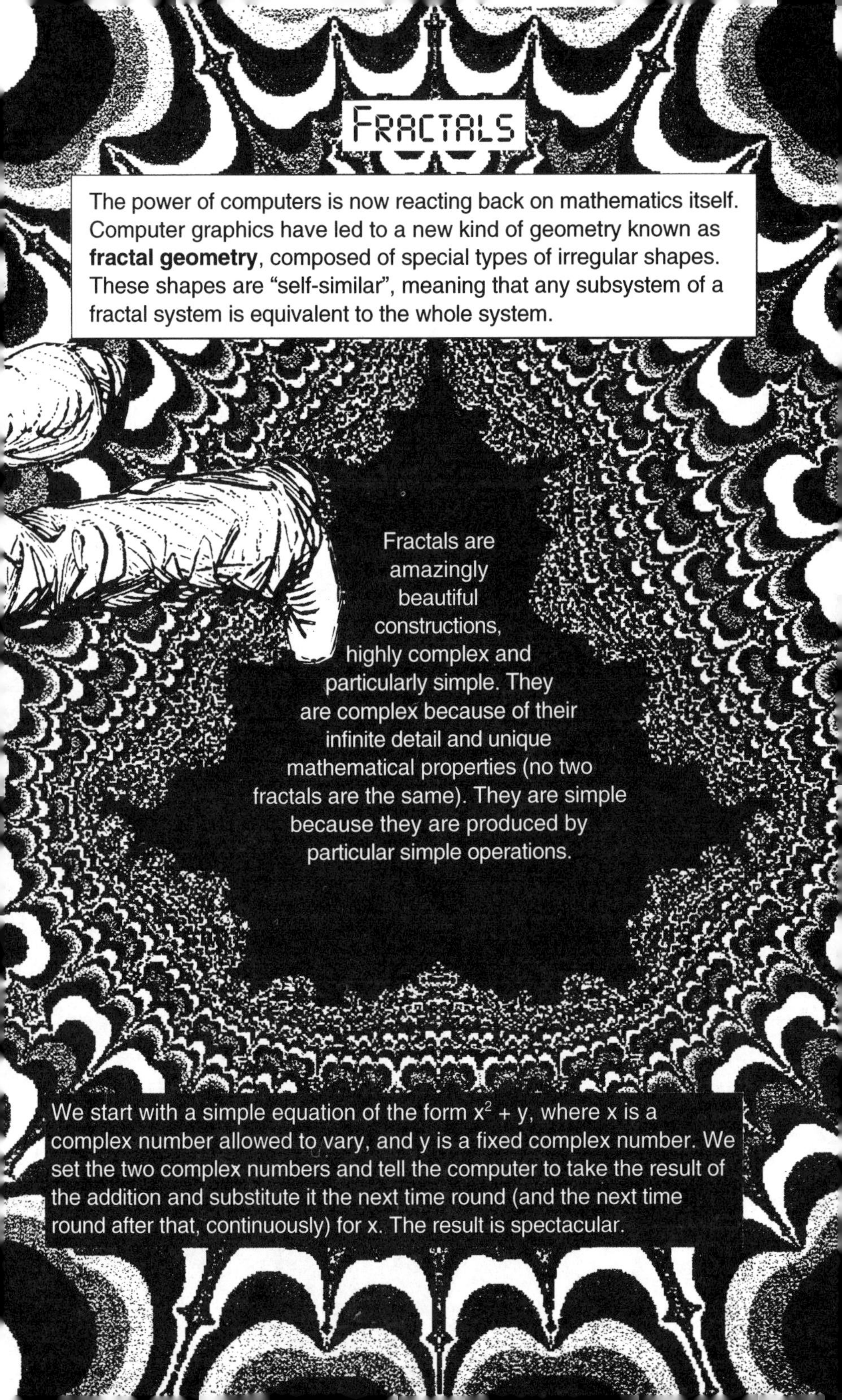

Fractals

The power of computers is now reacting back on mathematics itself. Computer graphics have led to a new kind of geometry known as **fractal geometry**, composed of special types of irregular shapes. These shapes are "self-similar", meaning that any subsystem of a fractal system is equivalent to the whole system.

Fractals are amazingly beautiful constructions, highly complex and particularly simple. They are complex because of their infinite detail and unique mathematical properties (no two fractals are the same). They are simple because they are produced by particular simple operations.

We start with a simple equation of the form $x^2 + y$, where x is a complex number allowed to vary, and y is a fixed complex number. We set the two complex numbers and tell the computer to take the result of the addition and substitute it the next time round (and the next time round after that, continuously) for x. The result is spectacular.

Benoit Mandelbrot (b. 1924), the Polish-born French mathematician who discovered fractals, described them as a way of seeing infinity.
My name is associated with the famous fractal on page 143, called the "Mandelbrot set".
Nowadays, fractals are used to study complex phenomena such as turbulence, distribution of earthquakes and evolution of cities.
And fractal geometry has led to the new mathematics of chaos theory.

Chaos Theory

Chaos theory describes phenomena which are not random, being described by differential equations, but which are not predictable either. This is because very slight changes in the initial conditions can produce large changes in the behaviour of the solutions. The classic statement (really, overstatement) of this property is that

...the fluttering of a butterfly's wings can change the course of a storm

Chaotic behaviour is closely tied in with the fractal property of systems. These are "self-similar", so that when we change the scale on which the behaviour is depicted, we see the same sort of variability. Apparently random phenomena, like the variation in prices on the stock markets, turn out to have this self-similar property. This makes it possible to use chaos theory

Perhaps the most important contribution of chaos theory to our understanding of mathematics is that it has made ignorance respectable.
For it has given mathematicians problems to solve which concern the impossibility of detailed knowledge.
The first time that the certainty of mathematics was definitely cracked, in the discovery of the paradoxes of the infinite around the turn of the 20th century, there was a sense of "crisis at the foundations".
This time, it is all part of progress; and in that way it reflects the continuing change in perception of what mathematics is all about.

Topology

The power of computers now reacts back on mathematics in other, more significant ways. Computers have produced proofs where human brainpower would have been inadequate. The most celebrated recent case is in the field of topology. Topology studies relationships among structures, independently of their precise shapes. It may be thought of as the mathematical field whose problems are the easiest to state and the hardest to solve.

One of the greatest challenges among topological problems is the "four-colour theorem". This states that any map can be coloured in with the use of at most four colours. The only rule is that no two countries with a common border can have the same colour. (It's alright if they meet at a point; otherwise the "map" could be a pie chart of as many sectors as we wish, requiring as many colours.) The only restriction is that each "country" is a single connected piece of land, and no country can have an "island" inside another (as happens with Italy and Switzerland near Lugano).

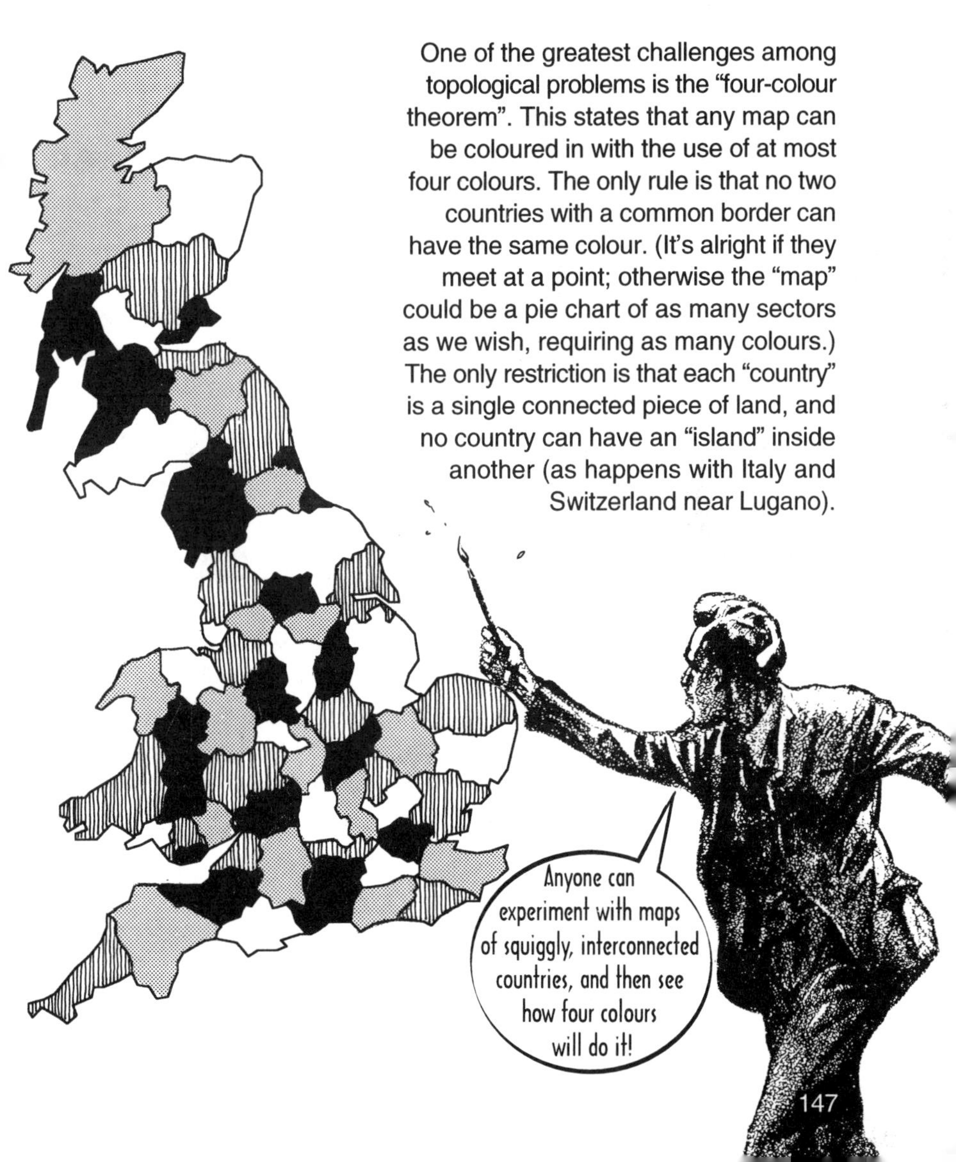
Anyone can experiment with maps of squiggly, interconnected countries, and then see how four colours will do it!

As mathematicians explored the four-colour problem, they discovered that the shape of the "world" is quite important.
On a "torus" (doughnut-shape), it was relatively easy to prove that five colours are sufficient.
But there is something peculiarly intractable about a sphere or a plane.

Eventually a proof was achieved in 1976. But it depended on the detailed study of more than a thousand cases, a task that was beyond human capabilities. So a computer program was created to test the special cases one at a time; and it worked, giving the desired result.

But then some mathematicians complained that they could not check the proof! For a computer program is not a sequence of logically connected statements – it is a set of instructions. Could we be sure that that particular program (unlike all others) had been debugged to absolute perfection? Eventually, a grumbling sort of consensus was achieved, and the proof is now accepted as "valid".

Number Theory

As in topology, problems in number theory are easy to describe and hard to prove.

For example, there is a "theorem" to the effect that every even number is the sum of two prime numbers.
So we try...
4 = 1 + 3,
6 = 3 + 3,
8 = 5 + 3,
16 = 11 + 5
Well, you carry on.

Proving this for all even numbers is quite difficult. Indeed, it was a challenge to mathematicians for a long time. The first successful attack on the problem, known as the "Goldbach conjecture", showed that no more than 400,000 primes are needed!

I've rounded up all the prime suspects, Mr 'Omes.
Show them in one-by-one, sergeant.

The most celebrated theorem in number theory is that of the French mathematician **Pierre de Fermat** (1601-65).

It comes out of my reflection on one of the very oldest among mathematical relationships, the "Pythagoras theorem", stating that there are "infinitely" many solutions of the equation...

$a^2 + b^2 = c^2$

where a, b and c are integers. The construction of such triplets of numbers has been known for centuries.

We have seen that Muslim mathematicians had thought about that relationship for higher powers. Some even tried to prove the impossibility of finding an example of numbers satisfying:

$x^3 + y^3 = z^3$

But Pierre Fermat thought he had done it, believing he had proved...

$x^n + y^n = z^n$

has no whole number solution for n greater than 2.

He actually wrote to a friend that he had a neat little proof of this, but it would not fit into the margin of the letter! So began the chase, which went on for three centuries and ended just recently. The proof is by the English mathematician **Andrew Wiles** (b. 1953) who now teaches at Princeton University.

It involves deeply abstruse mathematics and runs to thousands of lines, involving hundreds of calculations and logical links.

It all goes to show that the human mind can still achieve what computers fail to do!

Number theory has traditionally been one of the least applicable branches of mathematics. Yet as the progress of different fields develops, they interact in unexpected ways.

The science of "cryptography" (making and breaking codes) has traditionally been of interest only to soldiers and spies.
FIRE
Internet
Something must be done!

But it has suddenly become of great commercial, technological, and political importance, for the security of messages sent over the Internet depends completely on how difficult it is to crack their codes.

The best way to make a code is to use very large numbers, whose composition cannot easily be computed. Defining such numbers, and devising means of constructing and deconstructing them, involves the theory of numbers and of groups. So the most abstract mathematical fields can now find themselves at the cutting edge of practice. Since governments are very concerned with their ability to intercept and decipher any messages that might come from criminals or terrorists, the problem has become highly political as well.

Statistics

The most common point where mathematics impinges on the lives of ordinary citizens is in statistics. The term itself means “statecraft”, as when governments realize that they could do a better job if they had information on what is going on in the state. But just collecting huge masses of numbers is not enough; they must be aggregated, analysed and summarized in order to be useful.

In this work, the various measures of statistics are used, such as “average”. But any such number is only a representative of a collection; and while it reveals and clarifies the picture in some respects, it may well serve to conceal and obscure it in others.

To show how statistics works, let’s imagine a village where there are

a hundred peasants earning a pitiful $100 a year,

ten farmers earning a comfortable $1,000 a year,

and one landlord earning a handsome $10,000 a year.

Thus the "average" income is nearly three times that of most people!

The total income of the village is then $30,000, and divided by 111 households this gives a modest $270 per year.

Instead, we might take the "median" income (where just 50% have more), or the "mode" (the income held by most). In both cases this would be $100, which ignores those more fortunate. To provide a better picture of the income distribution, we might quote the lower and upper "deciles" (10% and 90% levels); the 90% decile would catch the eleventh household from the top, which is middle-income.

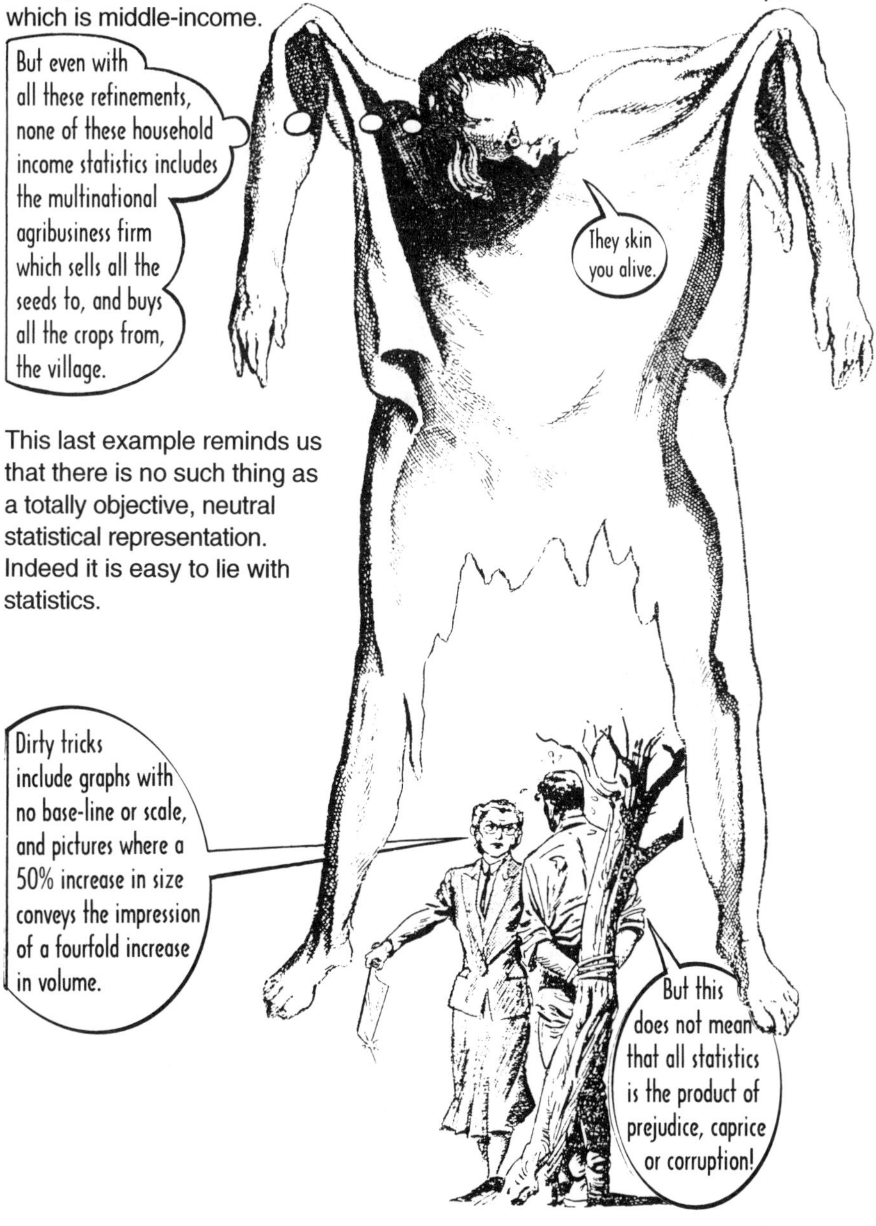

But even with all these refinements, none of these household income statistics includes the multinational agribusiness firm which sells all the seeds to, and buys all the crops from, the village.

They skin you alive.

This last example reminds us that there is no such thing as a totally objective, neutral statistical representation. Indeed it is easy to lie with statistics.

Dirty tricks include graphs with no base-line or scale, and pictures where a 50% increase in size conveys the impression of a fourfold increase in volume.

But this does not mean that all statistics is the product of prejudice, caprice or corruption!

P-values and Outliers

In all statistical tests of significance, there is quoted a number that is called a "confidence limit" or a "p-value". This might be 5%, 1%, or something else (or alternatively, 95%, 99%). Roughly speaking, this stands for the degree of certainty that the test conveys. It expresses the odds (20-1 or 100-1) against the test giving a false-positive result. No test can give perfect results! The higher the certainty required, the more expensive the test will be; and so those who set the standards for the particular field will have decided on the acceptable risk of each sort of possible error.

For a classic example, we have the sampling of apples, where a single rotten one can spoil a barrel...
...as contrasted to the sampling of artillery fuses, where one defective fuse can take the whole shipload with it.

There is a flip-side to these p-values, which are designed to limit the chances of false-positive results. A more rigorous p-value makes a test more "selective", but it also makes it less "sensitive". If we are testing for the toxicity of some environmental pollutant, our 95% p-value may protect us from false alarms, but it may then leave us vulnerable to false complacency. So, an apparently "objective" statistical significance test embodies an implicit "burden of proof": is the substance deemed safe until rigorously proved dangerous, or is an "early warning" sign to be admitted as valid? In each case, a "precautionary principle" is at work. The unavoidable question is, on whose behalf is the precaution applied?

Even in the simplest uses of statistics, as in the representation of experimental data, value-judgements are inevitable. Not all data hug closely to the line drawn through the points; indeed, if they are too close, that is a sign that they were fabricated. And some data points will be quite far from the crowd – we call them "outliers". If they are included in the calculation, they may bias it badly. But to reject them amounts to a judgement that there is something wrong with them, and that could amount to throwing away valuable, even crucial, information.

The first evidence of the "Ozone Hole" over Antarctica was missed for several years. Later it was discovered that it had been filtered out automatically, because the computer's statistical program deemed the data to be outliers.

Probability

The techniques for processing statistical data are mainly based on probability theory. This involves three quite distinct concepts, which are all too frequently confused.

First there is the "geometrical" probability based on symmetries, as when we say that the probability of throwing a seven with a pair of dice is just one-sixth.

(You can count that there are six ways to get a seven, out of thirty-six possibilities.)

Then there is the "empirical" probability of a person living beyond the age of seventy-five, which is based on statistics collected in the past.

And finally, there are the "judgements" of probability, such as the posted betting odds on the winner of a horse race or of a General Election.

Although they are conceptually distinct, the three sorts of probability are commonly used together without any clear distinction. Because of this, statistical inference has many pitfalls.

Suppose someone says to a friend:
This coin is biased – every time I toss it, I get Heads!
How many times have you tried it?

Once.
Oh, that's ridiculous. A single trial will always give "all" results the same.

So there is another toss, showing Heads again.
See? That proves it!
No, that's just a coincidence. That could occur any time. You must have more tosses than that.
OK! How many times must I get all Heads before I can say that the coin is biased?

Suddenly, the friend is perplexed. She knows that in an unbiased coin, Heads and Tails have equal geometrical probabilities. Because of that, "in the long run" an unbiased coin will tend to show an even number of Heads and Tails. That can be confirmed empirically. But to go from those two general facts, to making a judgement on whether the particular coin is biased, is another story altogether.

Judgements about whether a particular coin is biased or not require the mathematical theory of probability and statistics. The experimental design will then incorporate assumptions about the behaviour of the coin, along with the evaluation of error-costs and the setting of confidence limits for the final judgements. The coin-tossing incident has, when clarified, led us directly into some serious issues. While the direct form of the question is a simple statement of probabilities ("Heads and Tails equal for an unbiased coin"), the inverse form ("Is the coin biased?") involves judgements supported by statistical science.

When statistical arguments get tangled up with causality, the pitfalls are everywhere. There is a story about a man who would not travel by air...

I'm unwilling to take the chance of one-in-a-million that a bomb has been planted by a terrorist on the flight.
PSYCHIATRIST
A statistician told me that the probability of two bombs being on a single flight is only one-in-a-trillion...
But now you're a contented, frequent flyer. Why?
... and I'm carrying the first one!
TIK TIK TIK TIK
Blimey, Charlie!
AND THIS IS HOW MY STORY ENDS.
Your fallacy was not to see that carrying your own bomb had no effect on the intentions of possible terrorists. The probability of a second bomb being carried, conditional on the first being there, was just the same – one-in-a-million – as before.

Uncertainty

Those who have to provide numbers, either to policy makers or the public, have a cruel dilemma. If they give explanations of uncertainties, and their reservations about specific numbers, the result can be incomprehensible. But if they simplify and just provide a "magic number" that defines safety (frequently "one-in-a-million"), they can be accused of being misleading.

We experts want to be told that the lifetime population risk from a particular carcinogen is "between one-in-a-hundred-thousand and one-in-ten-million (at the 95% confidence level)".
But we, the public, want to know whether it is "safe", and if not, what precautions to take.
So communication of scientific results is far from simple and "objective".

The great challenge to mathematics on the social front is in the management of uncertainty. It had long been believed that the progress of natural science would push back ignorance and banish uncertainty – what remained could be tamed by probability theory.

But uncertainty has now come back to centre stage.
What are the chances of that happening, eh?
I'm not certain.

Uncertainty has conquered mathematics at its foundations, and it is at the core of the "quantum theory" of physics.

Now we are forced to confront the effects of our industrial civilization on an unpredictably complex natural environment. Uncertainty comes to the fore as never before. It is not surprising that popular new fields of mathematics are called "catastrophe" and "chaos". Now we can consider whether uncertainty should even be included in our ideas of what mathematics is all about.

Policy Numbers

Our understanding of numbers as devised for counting and calculating is not always appropriate for numbers used in policy-making. These uses require different conceptions and different skills. Because of our long tradition of focusing on mathematics as precise and true, we tend not to realize that uncertainty is an integral part of policy numbers. Excessive precision in numerical information in the media and official statements shrouds uncertainty in mystery. After all, if a quantity is expressed in two digits, as, say, 47, that means that it is different from 46 and 48, or is known to within about 2%.

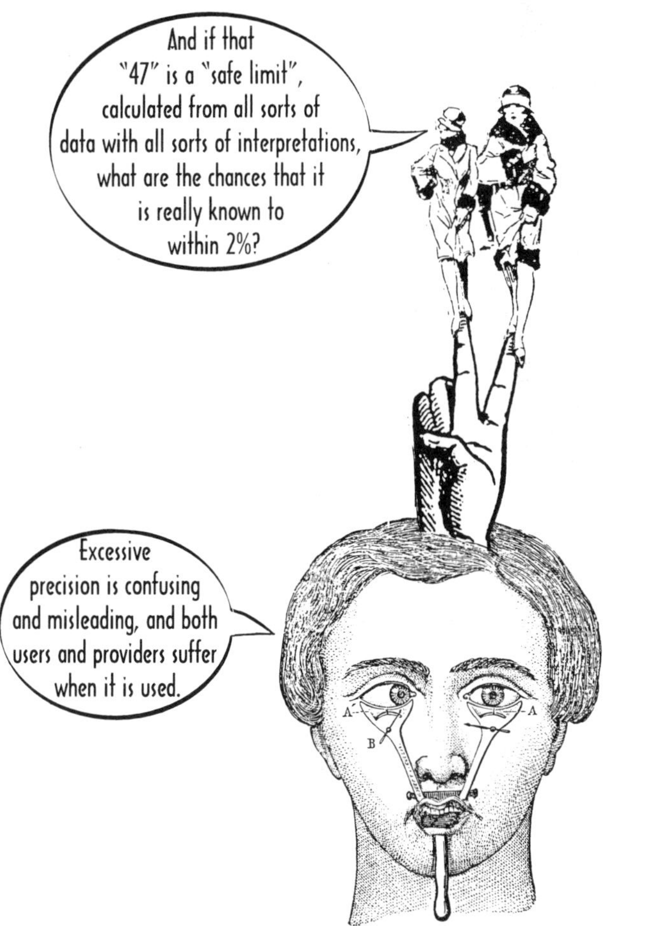
And if that "47" is a "safe limit", calculated from all sorts of data with all sorts of interpretations, what are the chances that it is really known to within 2%?
It's the work of the devil!
Excessive precision is confusing and misleading, and both users and providers suffer when it is used.
A
A
B

I will destroy the cities utterly because of their evil ways.
I plead with you, Lord, agree to spare the cities if just fifty righteous people can be found.
LOT'S WIFE
By the way, if there are just forty-five righteous, will you destroy the cities for lack of just five?
In the event, you could only locate Lot's family of four, and so the cities perished after all.

The significance of numbers in policy making depends on their context. There is a conversation in the Bible in which an astonishing sophistication is displayed. In Genesis 18, Abraham and the Lord are before the cities of Sodom and Gomorrah. The Lord says...

Then Abraham says...

So Abraham lifts the argument to another level. Now, it is not about **policy** (sparing the city if some righteous souls can be found), but about **implementation** (what if we are just a little bit under quota?). In this context, fifty is not a count but a policy-number, with an implied "fringe". Abraham was arguing that forty-five was within that fringe. Surely the Lord would not destroy the city because of a deficit of five, which in the context was below the limit of significance? The Lord gave in, on that estimate of the fringe. Perhaps sensing the skill of his adversary, he quickly allowed the quota to come down to ten righteous souls. Prudently, Abraham bargained no more.

The story of "saving Sodom" shows how numbers can have very different meanings in an argument. The "fifty" relates to the estimate, and the five, or forty-five, to its fringe. The difference between forty-five and fifty depends on the context. Sometimes the difference is significant (outside the fringe) and sometimes it is not. Although the example is about what we would now call a policy-number, the point about meaning depending on context holds for all estimates and measures.

The same sort of phenomenon can be seen in the "key-cutter's paradox". Someone starts with a new key that fits the lock, and then people take copies of copies. Each time, the copy is "exact" (within the allowable machine tolerances), but after repeated copying the new key does not fit. This is because the allowable errors in the copying machine have accumulated to the point where the multiply-copied key gets outside the tolerance range of the original lock-and-key fit. For that crucial dimension, we have (in terms of measurements): A = B = C = … = K. But A does not equal K. In terms of ordinary arithmetic, this is crazy. But it illustrates the point that numbers in estimations and measurements make sense only in specific contexts, and do not mean the same as in simple counting.

Mathematics and Eurocentrism

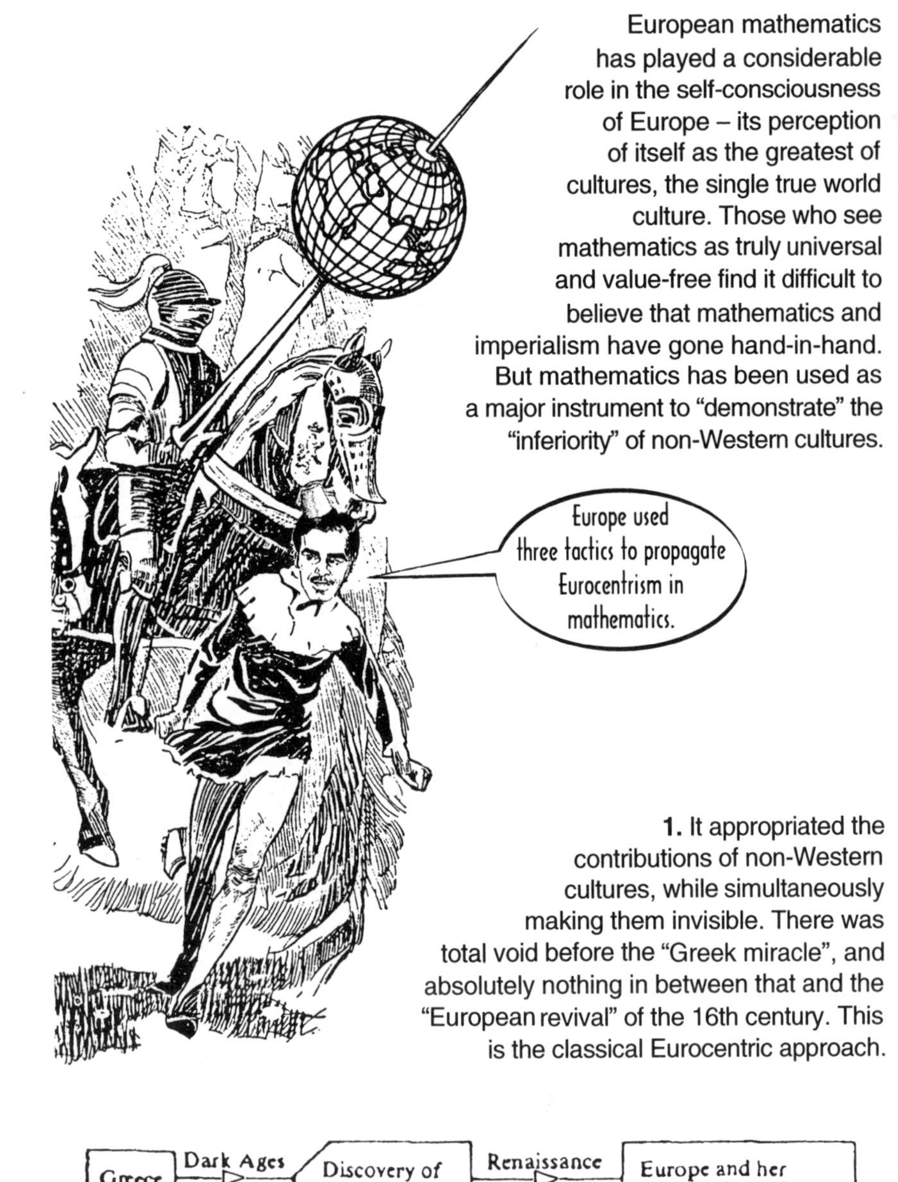
Europe used three tactics to propagate Eurocentrism in mathematics.

European mathematics has played a considerable role in the self-consciousness of Europe – its perception of itself as the greatest of cultures, the single true world culture. Those who see mathematics as truly universal and value-free find it difficult to believe that mathematics and imperialism have gone hand-in-hand. But mathematics has been used as a major instrument to "demonstrate" the "inferiority" of non-Western cultures.

1. It appropriated the contributions of non-Western cultures, while simultaneously making them invisible. There was total void before the "Greek miracle", and absolutely nothing in between that and the "European revival" of the 16th century. This is the classical Eurocentric approach.

Greece → Dark Ages → Discovery of Greek learning → Renaissance → Europe and her cultural dependencies

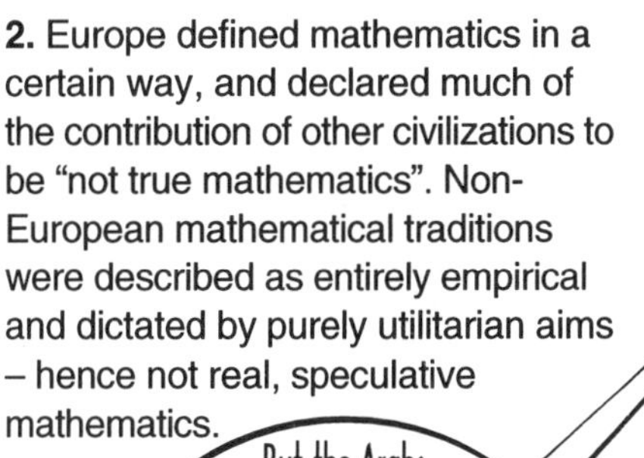

2. Europe defined mathematics in a certain way, and declared much of the contribution of other civilizations to be "not true mathematics". Non-European mathematical traditions were described as entirely empirical and dictated by purely utilitarian aims – hence not real, speculative mathematics.

But the Arabs were gracious enough to preserve the true Greek heritage of speculative mathematics and pass it on to its rightful inheritors, the European mathematicians of the Renaissance.

This is the "conveyor belt" theory of Eurocentrism.

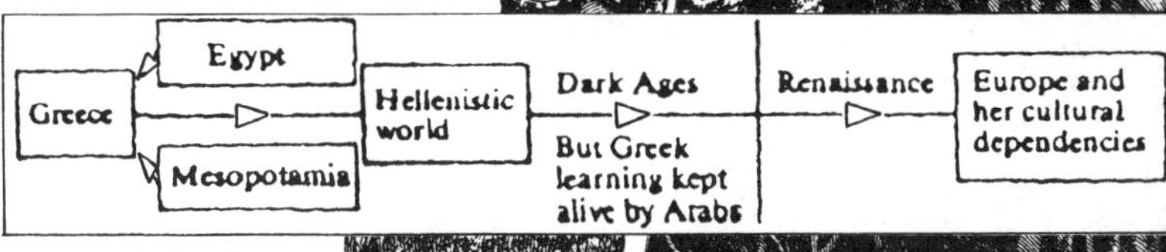

Greece
Egypt
Mesopotamia
Hellenistic world
Dark Ages
But Greek learning kept alive by Arabs
Renaissance
Europe and her cultural dependencies

3. It legitimized the "traditional" account of mathematical development as a purely European product, and institutionalized it in mathematical education.

So even today, throughout the world, mathematics is taught in terms of the ideology of imperialism.

The imperial experience prepared the students to consider it unthinkable that non-Europeans could produce mathematical knowledge. It fostered the myth that mathematics was a civilizing gift that Europe brought to the colonies, a Promethean spark that in time would enable the backward natives to penetrate the secrets of science and technology and enter the modern world.

George Gheverghese Joseph, British Asian historian of mathematics.

Ethnomathematics

At long last, "ethnomathematics" is being studied, promoted and used in teaching.
Ethnomathematics problematizes academic mathematics and brings onto the scene "Other" mathematics, usually not mentioned at school or university.

It seeks to establish a close relationship between mathematics, culture and society, and reminds us that "mathematics" includes more than the abstract theoretical studies of the Platonic tradition and the teaching curricula derived from them. We can see how much variety, ingenuity and creativity have gone into the ways in which different peoples accomplish and make sense of their mathematical tasks.

"Ethno" means "people", and ethnomathematics is the mathematics of all those people who have been excluded from knowledge and cultural production.
It includes mathematical traditions of non-Western civilizations such as those of China, India and Islam...
...as well as "vernacular" mathematics of ancient cultural traditions such as the "street mathematics" of peasant pushcart vendors in Brazil...
..."folk mathematics" of Latin American indigenous people...
...techniques of carpet-layers in America...
...even the mathematics involved in European women's knitting seen as algebra.
Just a minute! Where are the women in all this?
Turn the page and we'll see...

Thus, ethnomathematical practices include not only formal symbolic systems, but also spatial design, practical construction techniques, calculation methods, measurement in time and space, specific ways of reasoning and inferring, and other cognitive and material activities.

Mathematics and Gender

It is unfortunate but true that our mathematical heritage was created largely by "dead white males".

The few women who in past ages got a chance at distinction in mathematics are a curiosity. One of them, the French mathematician **Sophie Germain** (1776-1831), actually pretended to be a man in her correspondence with the German mathematician **Karl Friedrich Gauss** (1777-1855).

My secret was betrayed when Napoleon's army captured his city, Göttingen, and I used my influence to ensure his safety.
When the French commander presented the compliments of Mlle Germain to me, I was astonished; I had thought that my Paris correspondent was a young man!

Various causes have been suggested by psychologists for the traditional "inferiority" of women at mathematics.

But now that, on the whole, girls do better than boys in mathematics, this is seen as a social problem requiring urgent solution.

Where Now?

For more than a millennium, Western culture has been dominated by the Platonic vision of mathematics.

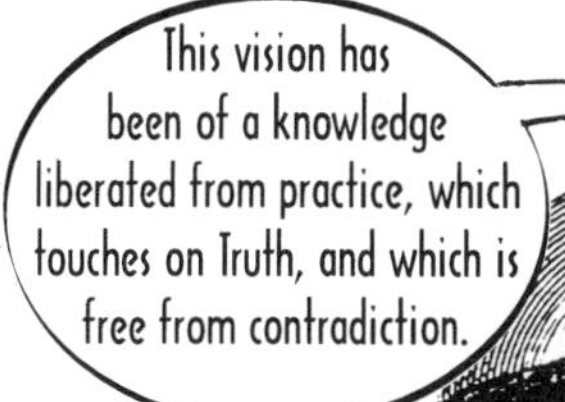
This vision has been of a knowledge liberated from practice, which touches on Truth, and which is free from contradiction.
The many discrepancies between vision and reality have been tucked away out of sight.

Philosophers, teachers, and popularizers alike have presented mathematics in that Platonic vision. Science has been imagined to be the application of mathematical truths. As part of the image, the contributions to mathematics from non-European cultures have been ignored or distorted.

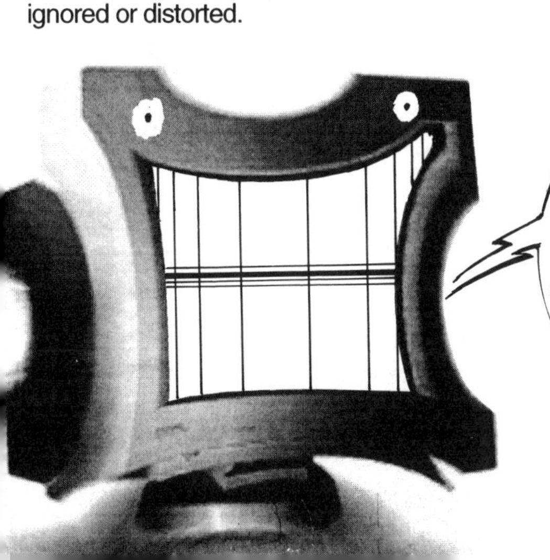
Even though mathematical research on "foundations" has destroyed the traditional certainties of mathematical thought, the rise of computers has brought "empirical" computational mathematics into a new synthesis with theory.

In spite of the broad literacy achieved by modern industrial society, effective numeracy is still restricted to a social and cultural élite.

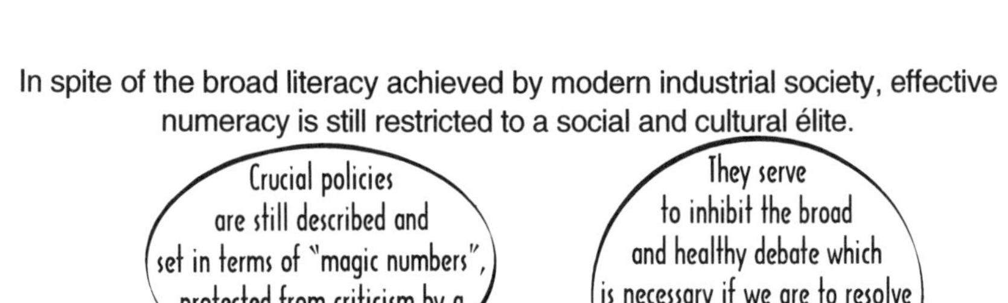
Crucial policies are still described and set in terms of "magic numbers", protected from criticism by a barrier of digits.
They serve to inhibit the broad and healthy debate which is necessary if we are to resolve the destructive contradictions of our industrial civilization.
FATIMA TURKISH CIGARETTES
Philip Davis
Reuben Hersh
There is hardly an area into which mathematics has not penetrated or might not penetrate. Just as all material objects, no matter where they are located, are subject to the law of gravity, so mathematics in its ability to deal with quantity, space, pattern, arrangement, structure, logical implication, has become, as Descartes would have wanted, the unifying glue of a rationalized world.
In days gone by, the ideas of intent, purpose, harmonies, imposed a reality on science that was derived from human values. Now, in the reverse direction, science, in its abstract mathematical formulations, has imposed its own reality on human values and behaviour.

Under these circumstances, it is essential for us all to know and appreciate the failure of mathematics (through science) to conquer the uncertainties of the practical world around us. It is necessary for us to think again about genuine knowledge and its achievement.

Mathematics therefore faces new challenges. And the citizen has a major role to play in meeting these challenges. In Bishop Berkeley's words, everyone...

...should use their own judgement, without a blind or mean deference to the best of mathematicians...
...on problems that are common to us all.
Conceiving new ways of living and of knowing, involving all peoples and all cultures, will require innovation in our social and scientific practices together.
In those, mathematics, finally liberated from its Eurocentric, Platonic image, will have a new role, with a new history of progress, new powers, and doubtless new paradoxes as well.

FURTHER READING

Read your Way
DYKE'S AUTOMOBILE & GASOLINE ENGINE ENCYCLOPEDIA
AUDELS CARPENTERS AND BUILDERS GUIDES

Books popularizing mathematics seemed to have increased exponentially; and sometimes it is not possible to pick out a good text from the plethora on offer. So for a "humanist" vision of mathematics through glimpses of its history, philosophy and practice, see P.J. Davis and R. Hersh, *The Mathematical Experience and Descartes' Dream* (Harvester, Brighton, 1981, 1986); for a monumental account of *Mathematics in Western Thought* (Penguin, London 1972) consult M. Kiline, who also provides the first systematic exposé of the suppressed conflicts about the foundations of mathematics in *Mathematics: The Loss of Certainty* (Oxford University Press, Oxford & New York 1980); and in his numerous books, Ian Stewart unravels the complexity of mathematics and makes it just that bit more enjoyable: begin with *Problems of Mathematics* (Oxford University Press, Oxford & New York 1987), and move on to *The Magical Maze* (Weidenfeld & Nicolson, London, 1998).

In *The Crest of the Peacock* (Penguin, London 1990), George G. Joseph reveals the "non-European roots of mathematics"; Donald Hill provides a readable account of Muslim mathematics in *Islamic Science and Engineering* (Edinburgh University Press, Edinburgh 1993); M. Ascher gives a "multicultural view of mathematical ideas" in *Ethnomathematics* (Brooks/Cole Publishing, Pacific Grove 1990); M.P. Closs (ed.) throws a spotlight on *Native American Mathematics* (University of Texas Press, Austin 1986); and Claudia Zaslavsky makes a pluralistic effort at relieving the *Fear of Maths* (Rutgers, New Jersey, 1994).

Simon Singh provides a riveting account of how *Fermat's Last Theorem* (Fourth Estate, London 1997) was recently proved; in *The Number Sense* (Allen Lane, London 1997), S. Dehaene explores the neuro-psychological approach to mathematical thinking; David Berlinski takes the reader on *A Tour of the Calculus* (Mandarin, London 1996); Ziauddin Sardar and Iwona Abrams provide a witty guide in *Introducing Chaos* (Icon Books, Cambridge 1998); and *Uncertainty and Quality in Science for Policy* by S.O. Funtowicz and J.R. Ravetz (Kluwer, Dordrecht 1990) is a pioneering look at policy numbers. Finally, *Mathematics for the Curious* by Peter Higgins (Oxford University Press, Oxford & New York 1998) satisfies, well, your curiosity.

The Authors

Ziauddin Sardar miscalculated and started his career as a physicist, but then moved on to become a science journalist and television reporter before settling as a writer and cultural critic. An internationally renowned thinker, his numerous publications include *Barbaric Others*, *Postmodernism and the Other* and *Cyberfutures*, which he co-edited with Jerry Ravetz. He has also written guides to Muhammad, Cultural Studies and Chaos in the *Introducing* series.

Jerry Ravetz is a "philosopher at large" of rare distinction. A Cambridge PhD in mathematics, he sits on the prestigious Working Party on the Production of a Public Understanding of Mathematics, and wrote the classic study, *Scientific Knowledge and Its Social Problems*. Formerly Reader in the History and Philosophy of Science at Leeds University, he has pioneered the study of uncertainty and policy numbers in social and scientific issues.

The Illustrator

Borin Van Loon. This is his seventh *Introducing* confection for Icon Books; earlier models are *Darwin* by Jonathan Miller, *Genetics* by Steve Jones, *Cultural Studies* by Zia Sardar, not to mention (though I will) *Buddha*, *Sociology,* and *Ancient Eastern Philosophy.* He is an author, illustrator and surrealist painter with a collage mural in London's Science Museum and stream-of-unconscious comic strips (which run the gamut from the Quantum's postulated multiple realities to the buttering of parsnips) coming out of his aristocratic and finely-chiselled ears.

INDEX

That's tall, folks!